Sociology 3U03

TABLE OF CONTENTS & ACKNOWLEDGEMENTS

PAGE

Labelling Theory   1
    Becker, H.
    <u>Constructions of Deviance: Social Power, Context, and Interaction</u>,
    Adler, P.A. and Adler, P. (eds.)
    © 2005 Cengage Learning Nelson Education
    This material has been copied under licence from Access Copyright.
    Resale or further copying of this material is strictly prohibited.

Primary and Secondary Deviation   5
    Lemert, E.
    <u>Social Deviance: Readings in Theory and Research</u>, Pontell, H.N. (ed.)
    © 2004 Prentice Hall Inc.
    This material has been copied under licence from Access Copyright.
    Resale or further copying of this material is strictly prohibited.

The Homosexual Role   7
    McIntosh, M.
    <u>Social Problems, 16.2</u>.
    © 1968 University of California Press
    This material has been copied under licence from Access Copyright.
    Resale or further copying of this material is strictly prohibited.

Gay Liberation and Lesbian Feminism   19
    Adam, B.D.
    <u>The Rise of a Gay and Lesbian Movement (Social Movements
    Past and Present)</u>, Adam, B.D.
    © 1987 Twayne Publishers
    This material has been copied under licence from Access Copyright.
    Resale or further copying of this material is strictly prohibited.

Homophobia and Women's Sports: The Disempowerment of Athletes   33
    Taub, D. E. and Blinde, E. M.
    <u>Constructions of Deviance: Social Power, Context, and Interaction</u>,
    Adler, P.A. and Adler, P. (eds.)
    © 2005 Cengage Learning Nelson Education
    This material has been copied under licence from Access Copyright.
    Resale or further copying of this material is strictly prohibited.

Masculinity as Homophobia: Fear, Shame, and Silence in the Construction of   41
Gender Identity
     Kimmel, M.
     The Gender of Desire: Essays on Male Sexuality, Kimmel, M.
     © 2005 State University of New York Press
     This material has been copied under licence from Access Copyright.
     Resale or further copying of this material is strictly prohibited.

The Five Sexes, Revisited   49
     Fausto-Sterling, A.
     The Sciences, 40.4
     © 2000 New York Academy of Sciences
     Reprinted with permission.

Am I Man Enough Yet? A Comparison of the Body Transition, Self Labeling,   57
and Sexual Orientation of Two Cohorts of Female-to-Male Transexuals
     Yerke, A.F. and Mitchell, V.
     International Journal of Transgenderism, 13.2
     © 2011 Taylor & Francis Inc. US
     Reprinted with permission.

Gender Expressions as a Reflection of Identity Reformation in Couple Partners   71
Following Disclosure of Male-to Female Transexualism
     Aramburu Alegria, C. and Ballard-Reisch, D.
     International Journal of Transgenderism, 14.2
     © 2013 Taylor & Francis Inc. US
     Reprinted with permission.

Seeking a Mate: Inter-Group Partnerships Among Gay Jewish Men   89
     Schnoor, R.F. and Weinfeld, M.
     Canadian Ethnic Studies Journal, 37.1
     © 2005 U of Calgary Press
     Reprinted with permission.

Negotiating Social Stigma Among Gay Asian Men   109
     Poon, M.K. and Ho, P.T.
     Sexualities, 11.1-2
     © 2008 Sage - Journals
     Reprinted with permission.

Modes of Suburban Gay Identity   133
     Brekhus, W.
     Deviance: The Interactionist Perspective, Rubington,
     E. and Weinberg, M.S. (eds.)
     © 2008 Allyn and Bacon - Pearson Education
     This material has been copied under licence from Access Copyright.
     Resale or further copying of this material is strictly prohibited.

Being Gay and Jewish: Negotiating Intersecting Identities   139
    Schnoor, R.F.
    <u>Sociology of Religion</u>, 67.1
    © 2006 Oxford University Press Journals
    Reprinted with permission.

Gay and Lesbian Christians: Homosexual and Religious Identity Integration in   157
the Members and Participants of a Gay-Positive Church
    Rodriguez, E.M. and Ouellette, S.C.
    <u>Journal for the Scientific Study of Religion</u>, 39.3
    © 2000 John Wiley & Sons (US)
    Reprinted with permission.

Identity Experience among Progressive Gay Muslims in North America: A   173
Qualitative Study within Al-Fatiha
    Minwalla, O.
    <u>Culture, Health & Sexuality</u>, 7.2
    © 2005 Taylor & Francis Inc. US
    Reprinted with permission.

7

Labeling Theory

HOWARD S. BECKER

Becker's classic statement of the labeling perspective agrees with Merton that the essence of crime is located outside of the individual person committing the deviant act. But rather than looking at the structural conditions that produce acts that might innately be regarded as criminal, Becker looks at societal reactions to behavior after the fact. Under other circumstances, the same behavior may be viewed very differently. Becker notes that variations may arise due to the diverse temporal contexts framing acts, to the social position and power of those who commit or who have been harmed by acts, and by the consequences that arise from acts. These framing elements, which are sometimes unrelated to the behavior itself, may elevate one act as heinous and relegate another, similar one, to obscurity. Becker thus locates the root of deviance in the response of other people rather than the act itself, and in the chain of events that is unleashed once people have labeled an act and its perpetrator as deviant.

The interactionist perspective ... defines deviance as the infraction of some agreed-upon rule. It then goes on to ask who breaks rules, and to search for the factors in their personalities and life situations that might account for the infractions. This assumes that those who have broken a rule constitute a homogeneous category, because they have committed the same deviant act.

Such an assumption seems to me to ignore the central fact about deviance: it is created by society. I do not mean this in the way it is ordinarily understood, in which the causes of deviance are located in the social situation of the deviant or in "social factors" which prompt his action. I mean, rather, that social groups create deviance by making the rules whose infraction constitutes deviance, and by applying those rules to particular people and labeling them as outsiders. From this point of view, deviance is *not* a quality of the act the person commits, but rather a consequence of the application by others of rules and sanctions to an "offender." The deviant is one to whom the label has successfully been applied; deviant behavior is behavior that people so label.[1]

Since deviance is, among other things, a consequence of the responses of others to a person's act, students of deviance cannot assume that they are dealing with a homogeneous category when they study people who have been labeled deviant. That is, they cannot assume that those people have actually committed a deviant act or broken some rule, because the process of labeling may not be infallible; some people may be labeled deviant who in fact have not broken a rule. Furthermore, they cannot assume that the category of those labeled deviant will contain all those who actually have broken a rule, for many offenders may escape apprehension and thus fail to be included in the population of "deviants" they study. Insofar as the category lacks homogeneity and fails to include all the cases that belong in it, one cannot reasonably expect to find common factors of personality or life situation that will account for the supposed deviance. What, then, do people who have been labeled deviant have in common? At the least, they share the label and the experience of being labeled as outsiders. I will begin my analysis with this basic similarity and view deviance as the product of a transaction that takes place between some social group and one who is viewed by that group as a rule-breaker. I will be less concerned with the personal and social characteristics of deviants than with the process by which they come to be thought of as outsiders and their reactions to that judgement....

The point is that the response of other people has to be regarded as problematic. Just because one has committed an infraction of a rule does not mean that others will respond as though this had happened. (Conversely, just because one has not violated a rule does not mean that he may not be treated, in some circumstances, as though he had.)

The degree to which other people will respond to a given act as deviant varies greatly. Several kinds of variation seem worth noting. First of all, there is variation over time. A person believed to have committed a given "deviant" act may at one time be responded to much more leniently than he would be at some other time. The occurrence of "drives" against various kinds of deviance illustrates this clearly. At various times, enforcement officials may decide to make an all-out attack on some particular kind of deviance, such as gambling, drug addiction, or homosexuality. It is obviously much more dangerous to engage in one of these activities when a drive is on than at any other time. (In a very interesting study of crime news in Colorado newspapers, Davis found that the amount of crime reported in Colorado newspapers showed very little association with actual changes in the amount of crime taking place in Colorado. And, further, that people's estimate of how much increase there had been in crime in Colorado was associated with the increase in the amount of crime news but not with any increase in the amount of crime.)[2]

The degree to which an act will be treated as deviant depends also on who commits the act and who feels he has been harmed by it. Rules tend to be applied more to some persons than others. Studies of juvenile delinquency make the point clearly. Boys from middle-class areas do not get as far in the legal process when they are apprehended as do boys from slum areas. The middle-class boy is less likely, when picked up by the police, to be taken to the station; less likely when taken to the station to be booked; and it is extremely unlikely that he will be convicted and sentenced.[3] This variation occurs even though the original infraction of the rule is the same in the two cases. Similarly, the law is differentially applied to Negroes and whites. It is well known that a Negro believed to have attacked a white woman is much more likely to be punished than a white man who commits the same offense; it is only slightly less well known that a Negro who murders another Negro is much less likely to be punished than a white man who commits murder.[4] This, of course, is one of the main points of Sutherland's analysis of white-collar crime: crimes committed by corporations are almost always prosecuted as civil cases, but the same crime committed by an individual is ordinarily treated as a criminal offense.[5]

Some rules are enforced only when they result in certain consequences. The unmarried mother furnishes a clear example. Vincent[6] points out that illicit sexual relations seldom result in severe punishment or social censure for the offenders. If, however, a girl becomes pregnant as a result of such activities the reaction of others is likely to be severe. (The illicit pregnancy is also an interesting example of the differential enforcement of rules on different categories of people. Vincent notes that unmarried fathers escape the severe censure visited on the mother.)

Why repeat these commonplace observations? Because, taken together, they support the proposition that deviance is not a simple quality, present in some kinds of behavior and absent in others. Rather, it is the product of a process which involves responses of other people to the behavior. The same behavior may be an infraction of the rules at one time and not at another; may be an infraction when committed by one person, but not when committed by another; some rules are broken with impunity, others are not. In short, whether a given act is deviant or not depends in part on the nature of the act (that is, whether or not it violates some rule) and in part on what other people do about it.

Some people may object that this is merely a terminological quibble, that one can, after all, define terms any way he wants to and that if some people want to speak of rule-breaking behavior as deviant without reference to the reactions of others they are free to do so. This, of course, is true. Yet it might be worthwhile to refer to such behavior as *rule-breaking behavior* and reserve the term *deviant* for those labeled as deviant by some segment of society. I do not insist that this usage be followed. But it should be clear that insofar as a scientist uses "deviant" to refer to any rule-breaking behavior and takes as his subject of study only those who have been *labeled* deviant, he will be hampered by the disparities between the two categories.

If we take as the object of our attention behavior which comes to be labeled as deviant, we must recognize that we cannot know whether a given act will be categorized as deviant until the response of others has occurred. Deviance is not a quality that lies in behavior itself, but in the interaction between the person who commits an act and those who respond to it....

In any case, being branded as deviant has important consequences for one's further social participation and self-image. The most important consequence is a drastic change in the individual's public identity. Committing the improper

vealed as a different kind of person from the kind he was supposed to be. He is labeled a "fairy," "dope fiend," "nut" or "lunatic," and treated accordingly.

In analyzing the consequences of assuming a deviant identity let us make use of Hughes' distinction between master and auxiliary status traits.[7] Hughes notes that most statuses have one key trait which serves to distinguish those who belong from those who do not. Thus the doctor, whatever else he may be, is a person who has a certificate stating that he has fulfilled certain requirements and is licensed to practice medicine; this is the master trait. As Hughes points out, in our society a doctor is also informally expected to have a number of auxiliary traits: most people expect him to be upper middle-class, white, male, and Protestant. When he is not, there is a sense that he has in some way failed to fill the bill. Similarly, though skin color is the master status trait determining who is Negro and who is white, Negroes are informally expected to have certain status traits and not to have others; people are surprised and find it anomalous if a Negro turns out to be a doctor or a college professor. People often have the master status trait but lack some of the auxiliary, informally expected characteristics; for example, one may be a doctor but be a female or a Negro.

Hughes deals with this phenomenon in regard to statuses that are well thought of, desired, and desirable (noting that one may have the formal qualifications for entry into a status but be denied full entry because of lack of the proper auxiliary traits), but the same process occurs in the case of deviant statuses. Possession of one deviant trait may have a generalized symbolic value, so that people automatically assume that its bearer possesses other undesirable traits allegedly associated with it.

To be labeled a criminal one need only commit a single criminal offense, and this is all the term formally refers to. Yet the word carries a number of connotations specifying auxiliary traits characteristic of anyone bearing the label. A man who has been convicted of housebreaking and thereby labeled criminal is presumed to be a person likely to break into other houses; the police, in rounding up known offenders for investigation after a crime has been committed, operate on this premise. Further, he is considered likely to commit other kinds of crimes as well, because he has shown himself to be a person without "respect for the law." Thus, apprehension for one deviant act exposes a person to the likelihood that he will be regarded as deviant or undesirable in other respects.

There is one other element in Hughes' analysis we can borrow with profit: the distinction between master and subordinate statuses.[8] Some statuses, in our society as in others, override all other statuses and have a certain priority. Race is one of these. Membership in the Negro race, as socially defined, will override most other status considerations in most other situations; the fact that one is a physician or middle-class or female will not protect one from being treated as a Negro first and any of these other things second. The status of deviant (depending on the kind of deviance) is this kind of master status. One receives the status as a result of breaking a rule, and the identification proves to be

NOTES

1. The most important earlier statements of this view can be found in Frank Tannenbaum, *Crime and the Community* (New York: Columbia University Press, 1938), and E. M. Lemert, *Social Pathology* (New York: McGraw-Hill Book Co., 1951). A recent article stating a position very similar to mine is John Kitsuse, "Societal Reaction to Deviance: Problems of Theory and Method," *Social Problems* 9 (Winter, 1962): 247–256.

2. F. James Davis, "Crime News in Colorado Newspapers," *American Journal of Sociology* LVII (January 1952): 325–330.

3. See Albert K. Cohen and James F. Short, Jr., "Juvenile Delinquency," p. 87 in Robert K. Merton and Robert A. Nisbet, eds., *Contemporary Social Problems* (New York: Harcourt, Brace and World, 1961).

4. See Harold Garfinkel, "Research Notes on Inter- and Intra-Racial Homicides," *Social Forces* 27 (May 1949): 369–381.

5. Edwin Sutherland, "White Collar Criminality," *American Sociological Review* V (February 1940): 1–12.

6. Clark Vincent, *Unmarried Mothers* (New York: The Free Press of Glencoe, 1961), pp. 3–5.

7. Everett C. Hughes, "Dilemmas and Contradictions of Status," *American Journal of Sociology* L (March 1945): 353–359.

8. *Ibid.*

Lemert, Edwin. "Primary and Secondary Deviation" in *Social Deviance: Readings in Theory and Research,* edited by Henry N. Pontell, pp. 74-77. © 2004 Prentice Hall.

Primary and Secondary Deviation*

EDWIN M. LEMERT

Sociopathic Individuation

The deviant person is a product of differentiating and isolating processes. Some persons are individually differentiated from others from the time of birth onward, as in the case of a child born with a congenital physical defect or repulsive appearance, and as in the case of a child born into a minority racial or cultural group. Other persons grow to maturity in a family or in a social class where pauperism, begging, or crime are more or less institutionalized ways of life for the entire group. In these latter instances the person's sociopsychological growth may be normal in every way, his status as a deviant being entirely caused by his maturation within the framework of social organization and culture designated as "pathological" by the larger society. This is true of many delinquent children in our society.[1]

It is a matter of great significance that the delinquent child, growing up in the delinquency areas of the city, has very little access to the cultural heritages of the larger conventional society. His infrequent contacts with this larger society are for the most part formal and external. Quite naturally his conception of moral values is shaped and molded by the moral code prevailing in his play groups and the local community in which he lives... the young delinquent has very little appreciation of the meaning of the traditions and formal laws of society.... Hence the conflict between the delinquent and the agencies of society is, in its broader aspects, a conflict of divergent cultures.

The same sort of gradual, unconscious process which operates in the socialization of the deviant child may also be recognized in the acquisition of socially unacceptable behavior by persons after having reached adulthood. However, with more

*From *Social Pathology: A Systematic Approach to the Theory of Sociopathic Behavior* by Edwin M. Lemert, pp. 75–78. Copyright 1951. Reprinted by permission of the author and McGraw-Hill, Inc.

verbal and sophisticated adults, step-by-step violations of societal norms tend to be progressively rationalized in the light of what is socially acceptable. Changes of this nature can take place at the level of either overt or covert behavior, but with a greater likelihood that adults will preface overt behavior changes with projective symbolic departures from society's norms. When the latter occur, the subsequent overt changes may appear to be "sudden" personality modifications. However, whether these changes are completely radical ones is to some extent a moot point. One writer holds strongly to the opinion that sudden and dramatic shifts in behavior from normal to abnormal are seldom the case, that a sequence of small preparatory transformations must be the prelude to such apparently sudden behavior changes. This writer is impressed by the day-by-day growth of "reserve potentialities" within personalities of all individuals, and he contends that many normal persons carry potentialities for abnormal behavior, which, given proper conditions, can easily be called into play.[2]

Personality Changes Not Always Gradual

This argument is admittedly sound for most cases, but it must be taken into consideration that traumatic experiences often speed up changes in personality.[3] Nor can the "trauma" in these experiences universally be attributed to the unique way in which the person conceives of the experience subjectively. Cases exist to show that personality modifications can be telescoped or that there can be an acceleration of such changes caused largely by the intensity and variety of the social stimulation. Most soldiers undoubtedly have entirely different conceptions of their roles after intensive combat experience. Many admit to having "lived a lifetime" in a relatively short period of time after they have been under heavy fire in battle for the first time. Many generals have remarked that their men have to be a little "shooted" or "blooded" in order to become good soldiers. In the process of group formation, crises and interactional amplification are vital requisites to forging true, role-oriented group behavior out of individuated behavior.[4]

The importance of the person's conscious symbolic reactions to his or her own behavior cannot be overstressed in explaining the shift from normal to abnormal behavior or from one type of pathological behavior to another, particularly where behavior variations become systematized or structured into pathological roles. This is not to say that conscious choice is a determining factor in the differentiating process. Nor does it mean that the awareness of the self is a purely conscious perception. Much of the process of self-perception is doubtless marginal from the point of view of consciousness.[5] But however it may be perceived, the individual's self-definition is closely connected with such things as self-acceptance, the subordination of minor to major roles, and with the motivation involved in learning the skills, techniques, and values of a new role. *Self-definitions or self-realizations are likely to be the result of sudden perceptions and they are especially significant when they are followed immediately by overt demonstrations of the new role they symbolize.* The self-defining junctures are critical points of personality genesis and in the special case of the atypical person they mark a division between two different types of deviation.

Primary and Secondary Deviation

There has been an embarrassingly large number of theories, often without any relationship to a general theory, advanced to account for various specific pathologies in human behavior. For certain types of pathology, such as alcoholism, crime, or stuttering, there are almost as many theories as there are writers on these subjects. This has been occasioned in no small way by the preoccupation with the origins of pathological behavior and by the fallacy of confusing *original* causes with *effective* causes. All such theories have elements of truth, and the divergent viewpoints they contain can be reconciled with the general theory here if it is granted that original causes or antecedents of deviant behaviors are many and diversified. This holds especially for the psychological processes leading to similar pathological behavior, but it also holds for the situational concomitants of the initial aberrant conduct. A person may come to use excessive alcohol not only for a wide variety of subjective reasons but also because of diversified situational influences, such as the death of a loved one, business failure, or participating in some sort of organized group activity calling for heavy drinking of liquor. Whatever the original reasons for violating the norms of the community, they are important only for certain research purposes, such as assessing the extent of the "social problem" at a given time or determining the requirements for a rational program of social control. From a narrower sociological viewpoint the deviations are not significant until they are organized subjectively and transformed into active roles and become the social criteria for assigning status. The deviant individuals must react symbolically to their own behavior aberrations and fix them in their socio-psychological patterns. The deviations remain primary deviations or symptomatic and situational as long as they are rationalized or otherwise dealt with as functions of a socially acceptable role. Under such conditions normal and pathological behaviors remain strange and somewhat tensional bedfellows in the same person. Undeniably a vast amount of such segmental and partially integrated pathological behavior exists in our society and has impressed many writers in the field of social pathology.

Just how far and for how long a person may go in dissociating his sociopathic tendencies so that they are merely troublesome adjuncts of normally conceived roles is not known. Perhaps it depends upon the number of alternative definitions of the same overt behavior that he can develop; perhaps certain physiological factors (limits) are also involved. However, if the deviant acts are repetitive and have a high visibility, and if there is a severe societal reaction, which, through a process of identification is incorporated as part of the "me" of the individual, the probability is greatly increased that the integration of existing roles will be disrupted and that reorganization based upon a new role or roles will occur. (The "me" in this context is simply the subjective aspect of the societal reaction.) Reorganization may be the adoption of another normal role in which the tendencies previously defined as "pathological" are given a more acceptable social expression. The other general possibility is the assumption of a deviant role, if such exists; or, more rarely, the person may organize an aberrant sect or group in which he creates a special role of his own. *When a person begins to employ his deviant behavior or a role based upon it as a means of defense, attack, or adjustment to the overt and covert problems created by the consequent societal reaction to him, his deviation is secondary.* Objective evidences of this change will be found in the symbolic appurtenances of the new role, in clothes, speech, posture, and mannerisms, which in some cases heighten social visibility, and which in some cases serve as symbolic cues to professionalization.

Role Conceptions of the Individual Must Be Reinforced by Reactions of Others

It is seldom that one deviant act will provoke a sufficiently strong societal reaction to bring about secondary deviation, unless in the process of introjection the individual imputes or projects meanings into the social situation which are not present. In this case anticipatory fears are involved. For example, in a culture where a child is taught sharp distinctions between "good" women and "bad" women, a single act of questionable morality might conceivably have a profound meaning for the girl so indulging. However, in the absence of reactions by the person's family, neighbors, or the larger community, reinforcing the tentative "bad-girl" self-definition, it is questionable whether a transition to secondary deviation would take place. It is also doubtful whether a temporary exposure to a severe punitive reaction by the community will lead a person to identify himself with a pathological role, unless, as we have said, the experience is highly traumatic. Most frequently there is a progressive reciprocal relationship between the deviation of the individual and the societal reaction, with a compounding of the societal reaction out of the minute accretions in the deviant behavior, until a point is reached where in-grouping and out-grouping between society and the deviant is manifest.[6] At this point a stigmatizing of the deviant occurs in the form of name calling, labeling, or stereotyping.

The sequence of interaction leading to secondary deviation is roughly as follows: (1) primary deviation; (2) social penalties; (3) further primary deviation; (4) stronger penalties and rejections; (5) further deviation, perhaps with hostilities and resentment beginning to focus upon those doing the penalizing; (6) crisis reached in the tolerance quotient, expressed in formal action by the community stigmatizing of the deviant; (7) strengthening of the deviant conduct as a reaction to the stigmatizing and penalties; (8) ultimate acceptance of deviant social status and efforts at adjustment on the basis of the associated role.

As an illustration of this sequence the behavior of an errant schoolboy can be cited. For one reason or another, let us say excessive energy, the schoolboy engages in a classroom prank. He is penalized for it by the teacher. Later, due to clumsiness, he creates another disturbance and again he is reprimanded. Then, as something happens, the boy is blamed for something he did not do. When the teacher uses the tag "bad boy" or "mischief maker" or other invidious terms, hostility and resentment are excited in the boy, and he may feel that he is blocked in playing the role expected of him. Thereafter, there may be a strong temptation to assume his role in the class as defined by the teacher, particularly when he discovers that there are rewards as well as penalties deriving from such a role. There is, of course, no implication here that such boys go on to become delinquents or criminals, for the mischief-maker role may later become integrated with or retrospectively rationalized as part of a role more acceptable to school authorities.[7] If such a boy continues this unacceptable role and becomes delinquent, the process must be accounted for in the light of the general theory of this volume. There must be a spreading corroboration of a sociopathic self-conception and societal reinforcement at each step in the process.

The most significant personality changes are manifest when societal definitions and their subjective counterpart become generalized. When this happens, the range of major role choices becomes narrowed to one general class.[8] This was very obvious in the case of a young girl who was the daughter of a paroled convict and who was attending a small Middle Western college. She continually argued with herself and with the author, in whom she had confided, that in reality she belonged on the "other side of the railroad tracks" and that her life could be enormously simplified by acquiescing in this verdict and living accordingly. While in her case there was a tendency to dramatize her conflicts, nevertheless there was enough societal reinforcement of her self-conception by the treatment she received in her relationship with her father and on dates with college boys to lend it a painful reality. Once these boys took her home to the shoddy dwelling in a slum area where she lived with her father, who was often in a drunken condition, they abruptly stopped seeing her again or else became sexually presumptive.

Notes

1. Shaw, C., *The Natural History of a Delinquent Career*, Chicago, 1941, pp. 75–76. Quoted by permission of the University of Chicago Press, Chicago.
2. Brown, L. Guy, *Social Pathology*, 1942, pp. 44–45.
3. Allport, G., *Personality, A Psychological Interpretation*, 1947, p. 57.
4. Slavson, S. R., *An Introduction to Group Psychotherapy*, 1943, pp. 10, 229ff.
5. Murphy, G., *Personality*, 1947, p. 482.
6. Mead, G., "The Psychology of Punitive Justice," *American Journal of Sociology*, 23 March, 1918, pp. 577–602.
7. Evidence for fixed or inevitable sequences from predelinquency to crime is absent. Sutherland, E. H., *Principles of Criminology*, 1939, 4th ed., p. 202.
8. Sutherland seems to say something of this sort in connection with the development of criminal behavior. *Ibid.*, p. 86.

THE HOMOSEXUAL ROLE

MARY McINTOSH
University of Leicester, England

> The current conceptualization of homosexuality as a condition is a false one, resulting from ethnocentric bias. Homosexuality should be seen rather as a social role. Anthropological evidence shows that the role does not exist in all societies, and where it does it is not always the same as in modern western societies. Historical evidence shows that the role did not emerge in England until towards the end of the seventeenth century. Evidence from the "Kinsey Reports" shows that, in spite of the existence of the role in our society, much homosexual behavior occurs outside the recognized role and the polarization between the heterosexual man and the homosexual man is far from complete.

Recent advances in the sociology of deviant behavior have not yet affected the study of homosexuality, which is still commonly seen as a condition characterizing certain persons in the way that birthplace or deformity might characterize them. The limitations of this view can best be understood if we examine some of its implications. In the first place, if homosexuality is a condition, then people either have it or do not have it. Many scientists and ordinary people assume that there are two kinds of people in the world: homosexuals and heterosexuals. Some of them recognize that homosexual feelings and behavior are not confined to the persons they would like to call "homosexuals" and that some of these persons do not actually engage in homosexual behavior. This should pose a crucial problem; but they evade the crux by retaining their assumption and puzzling over the question of how to tell whether someone is "really" homosexual or not. Lay people too will discuss whether a certain person is "queer" in much the same way as they might question whether a certain pain indicated cancer. And in much the same way they will often turn to scientists or to medical men for a surer diagnosis. The scientists, for their part, feel it incumbent on them to seek criteria for diagnosis.

Thus one psychiatrist, discussing the definition of homosexuality, has written:

> ... I do not diagnose patients as homosexual unless they have engaged in overt homosexual behavior. Those who also engage in heterosexual activity are diagnosed as bisexual. An isolated experience may not warrant the diagnosis, but repetetive (sic) homosexual behavior in adulthood, whether sporadic or continuous, designates a homosexual.[1]

Along with many other writers, he introduces the notion of a third type of person, the "bisexual," to handle the fact that behavior patterns cannot be conveniently dichotomized into hetero-

---

[1] Irving Bieber, "Clinical Aspects of Male Homosexuality," in Judd Marmor, editor, *Sexual Inversion*, New York: Basic Books, 1965, p. 248; this is but one example among many.

sexual and homosexual. But this does not solve the conceptual problem, since bisexuality too is seen as a condition (unless as a passing response to unusual situations such as confinement in a one-sex prison). In any case there is no extended discussion of bisexuality; the topic is usually given a brief mention in order to clear the ground for the consideration of "true homosexuality."

To cover the cases where the symptoms of behavior or of felt attractions do not match the diagnosis, other writers have referred to an adolescent homosexual phase or have used such terms as "latent homosexual" or "pseudo homosexual." Indeed one of the earliest studies of the subject, by Krafft-Ebing, was concerned with making a distinction between the "invert" who is congenitally homosexual and others who, although they behave in the same way, are not true inverts.[2]

A second result of the conceptualization of homosexuality as a condition is that the major research task has been seen as the study of its etiology. There has been much debate as to whether the condition is innate or acquired. The first step in such research has commonly been to find a sample of "homosexuals" in the same way that a medical researcher might find a sample of diabetics if he wanted to study that disease. Yet, after a long history of such studies, the results are sadly inconclusive and the answer is still as much a matter of opinion as it was when Havelock Ellis published *Sexual Inversion*[3] seventy years ago. The failure of research to answer the question has not been due to lack of scientific rigor or to any inadequacy of the available evidence; it results rather from the fact that the wrong question has been asked. One might as well try to trace the etiology of "committee-chairmanship" or "Seventh-Day Adventism" as of "homosexuality."

The vantage-point of comparative sociology enables us to see that the conception of homosexuality as a condition is, in itself, a possible object of study. This conception and the behavior it supports operate as a form of social control in a society in which homosexuality is condemned. Furthermore, the uncritical acceptance of the conception by social scientists can be traced to their concern with homosexuality as a social problem. They have tended to accept the popular definition of what the problem is and they have been implicated in the process of social control.

The practice of the social labeling of persons as deviant operates in two ways as a mechanism of social control.[4] In the first place it helps to provide a clear-cut, publicized, and recognizable threshold between permissible and impermissible behavior. This means that people cannot so easily drift into deviant behavior. Their first moves in a deviant direction immediately raise the question of a total move into a deviant role with all the sanctions that this is likely to elicit. Secondly, the labeling serves to segregate the deviants from others and this means that their deviant practices and their self-justifications for these practices are contained within a relatively narrow group. The creation

---

[2] R. von Krafft-Ebing, *Psychopathia Sexualis*, 1889.

[3] Later published in H. Ellis, *Studies in the Psychology of Sex*, Vol. 2, New York: Random House, 1936.

[4] This is a grossly simplified account. Edwin Lemert provides a far more subtle and detailed analysis in *Social Pathology*, New York: McGraw-Hill, 1951, ch. 4, "Sociopathic Individuation."

of a specialized, despised, and punished role of homosexual keeps the bulk of society pure in rather the same way that the similar treatment of some kinds of criminals helps keep the rest of society law-abiding.

However, the disadvantage of this practice as a technique of social control is that there may be a tendency for people to become fixed in their deviance once they have become labeled. This, too, is a process that has become well-recognized in discussions of other forms of deviant behavior such as juvenile delinquency and drug taking and, indeed, of other kinds of social labeling such as streaming in schools and racial distinctions. One might expect social categorizations of this sort to be to some extent self-fulfilling prophecies: if the culture defines people as falling into distinct types—black and white, criminal and non-criminal, homosexual and normal—then these types will tend to become polarized, highly differentiated from each other. Later in this paper I shall discuss whether this is so in the case of homosexuals and "normals" in the United States today.

It is interesting to notice that homosexuals themselves welcome and support the notion that homosexuality is a condition. For just as the rigid categorization deters people from drifting into deviancy, so it appears to foreclose on the possibility of drifting back into normality and thus removes the element of anxious choice. It appears to justify the deviant behavior of the homosexual as being appropriate for him as a member of the homosexual category. The deviancy can thus be seen as legitimate for him and he can continue in it without rejecting the norms of the society.[5]

The way in which people become labeled as homosexual can now be seen as an important social process connected with mechanisms of social control. It is important, therefore, that sociologists should examine this process objectively and not lend themselves to participation in it, particularly since, as we have seen, psychologists and psychiatrists on the whole have not retained their objectivity but become involved as diagnostic agents in the process of social labeling.[6]

It is proposed that the homosexual should be seen as playing a social role rather than as having a condition. The role of "homosexual," however, does not simply describe a sexual behavior pattern. If it did, the idea of a role would be no more useful than that of a condition. For the purpose of introducing the term "role" is to enable us to handle the fact that behavior in this sphere does not match popular beliefs: that sexual behavior patterns cannot be dichotomized in the way that the social roles of homosexual and heterosexual can.

It may seem rather odd to distinguish in this way between role and behavior, but if we accept a definition of role in terms of expectations (which may or may not be fulfilled), then the distinction is both legitimate and useful. In modern societies where a separate homosexual role is recognized, the expectation, on behalf of those who play the role and of others, is that a homosexual will be exclusively or very predominantly homosexual in his feelings and

---

[5] For discussion of situations in which deviants can lay claim to legitimacy, see Talcott Parsons, *The Social System*, New York: Free Press, 1951, pp. 292-293.

[6] The position taken here is similar to that of Erving Goffman in his discussion of becoming a mental patient; *Asylums*, Garden City, N.Y.: Doubleday-Anchor, 1961, pp. 128-146.

behavior. In addition, there are other expectations that frequently exist, especially on the part of nonhomosexuals, but affecting the self-conception of anyone who sees himself as homosexual. These are: the expectation that he will be effeminate in manner, personality, or preferred sexual activity; the expectation that sexuality will play a part of some kind in all his relations with other men; and the expectation that he will be attracted to boys and very young men and probably willing to seduce them. The existence of a social expectation, of course, commonly helps to produce its own fulfillment. But the question of how far it is fulfilled is a matter for empirical investigation rather than *a priori* pronouncement. Some of the empirical evidence about the chief expectation—that homosexuality precludes heterosexuality—in relation to the homosexual role in America is examined in the final section of this paper.[7]

In order to clarify the nature of the role and demonstrate that it exists only in certain societies, we shall present the cross-cultural and historical evidence available. This raises awkward problems of method because the material has hitherto usually been collected and analyzed in terms of culturally specific modern western conceptions.

THE HOMOSEXUAL ROLE IN VARIOUS SOCIETIES

To study homosexuality in the past or in other societies we usually have to rely on secondary evidence rather than on direct observation. The reliability and the validity of such evidence is open to question because what the original observers reported may have been distorted by their disapproval of homosexuality and by their definition of it, which may be different from the one we wish to adopt.

For example, Marc Daniel tries to refute accusations of homosexuality against Pope Julian II by producing four arguments: the Pope had many enemies who might wish to blacken his name; he and his supposed lover, Alidosi, both had mistresses; neither of them was at all effeminate; and the Pope had other men friends about whom no similar accusations were made.[8] In other words Daniel is trying to fit an early sixteenth century Pope to the modern conception of the homosexual as effeminate, exclusively homosexual, and sexual in relation to all men. The fact that he does not fit is, of course, no evidence, as Daniel would have it, that his relationship with Alidosi was not a sexual one.

Anthropologists too can fall into this trap. Marvin Opler, summarizing anthropological evidence on the subject, says,

> Actually, no society, save perhaps Ancient Greece, pre-Meiji Japan, certain top echelons in Nazi Germany, and the scattered examples of such special status groups as the berdaches, Nata slaves, and one category of Chuckchee shamans, has lent sanction in any real sense to homosexuality.[9]

Yet he goes on to discuss societies in which there are reports of sanctioned adolescent and other occasional "experimentation." Of the Cubeo of the

---

[7] For evidence that many self-confessed homosexuals in England are not effeminate and many are not interested in boys, see Michael Schofield, *Sociological Aspects of Homosexuality*, London: Longmans, 1965.

[8] Marc Daniel, "Essai de méthodologie pour l'étude des aspects homosexuels de l'histoire," *Arcadie*, 133 (January, 1965), pp. 31-37.

[9] Marvin Opler, "Anthropological and Cross-Cultural Aspects of Homosexuality," in Marmor, editor, *op. cit.*, p. 174.

North West Amazon, for instance, he says, "*true* homosexuality among the Cubeo is rare if not absent," giving as evidence the fact that no males with persistent homosexual patterns are reported.[10]

Allowing for such weaknesses, the Human Relations Area Files are the best single source of comparative information. Their evidence on homosexuality has been summarized by Ford and Beach,[11] who identify two broad types of accepted patterns: the institutionalized homosexual role and the liaison between men or boys who are otherwise heterosexual.

The recognition of a distinct role of *berdache* or transvestite is, they say, "the commonest form of institutionalized homosexuality." This form shows a marked similarity to that in our own society, though in some ways it is even more extreme. The Mohave Indians of California and Arizona, for example,[12] recognized both an *alyhā*, a male transvestite who took the role of the woman in sexual intercourse, and a *hwamē*, a female homosexual who took the role of the male. People were believed to be born as *alyhā* or *hwamē*, hints of their future proclivities occurring in their mothers' dreams during pregnancy. If a young boy began to behave like a girl and take an interest in women's things instead of men's, there was an initiation ceremony in which he would become an *alyhā*. After that he would dress and act like a woman, would be referred to as "she" and could take "husbands."

But the Mohave pattern differs from ours in that although the *alyhā* was considered regrettable and amusing, he was not condemned and was given public recognition. The attitude was that "he was an *alyhā*, he could not help it." But the "husband" of an *alyhā* was an ordinary man who happened to have chosen an *alyhā*, perhaps because they were good housekeepers or because they were believed to be "lucky in love," and he would be the butt of endless teasing and joking.

This radical distinction between the feminine passive homosexual and his masculine active partner is one which is not made very much in our own society,[13] but which is very important in the Middle East. There, however, neither is thought of as being a "born" homosexual, although the passive partner, who demeans himself by his feminine submission, is despised and ridiculed, while the active one is not. In most of the ancient Middle East, including among the Jews until the return from the Babylonian exile, there were male temple prostitutes.[14] Thus even cultures that recognize a separate homosexual role may not define it in the same way as our culture does.

Many other societies accept or approve of homosexual liaisons as part of a variegated sexual pattern. Usually

---

[10] *Ibid.*, p. 117.

[11] C. S. Ford and F. A. Beach, *Patterns of Sexual Behavior*, New York: Harper, 1951, ch. 7.

[12] George Devereux, "Institutionalized Homosexuality of the Mohave Indians," *Human Biology*, Vol. 9, 1937, pp. 498-527; reprinted in Hendrik M. Ruitenbeek, editor, *The Problem of Homosexuality in Modern Society*, New York: Dutton, 1963.

[13] The lack of cultural distinction is reflected in behavior; Gordon Westwood found that only a small proportion of his sample of British homosexuals engaged in anal intercourse and many of these had been both active and passive and did not have a clear preference. See *A Minority*, London: Longmans, 1960, pp. 127-134.

[14] Gordan Rattray Taylor, "Historical and Mythological Aspects of Homosexuality," in Marmor, *op. cit.*; Fernando Henriques, *Prostitution and Society*, Vol. 1, London: MacGibbon and Kee, 1962, pp. 341-343.

these are confined to a particular stage in the individual's life. Among the Aranda of Central Australia, for instance, there are long-standing relationships of several years' duration, between unmarried men and young boys, starting at the age of ten to twelve.[15] This is rather similar to the well-known situation in classical Greece, but there, of course, the older man could have a wife as well. Sometimes, however, as among the Siwans of North Africa,[16] all men and boys can and are expected to engage in homosexual activities, apparently at every stage of life. In all of these societies there may be much homosexual behavior, but there are no "homosexuals."

THE DEVELOPMENT OF THE HOMOSEXUAL ROLE IN ENGLAND

The problem of method is even more acute in dealing with historical material than with anthropological, for history is usually concerned with "great events" rather than with recurrent patterns. There are some records of attempts to curb sodomy among minor churchmen during the medieval period,[17] which seem to indicate that it was common. At least they suggest that laymen feared on behalf of their sons that it was common. The term "catamite" meaning "boy kept for immoral purposes," was first used in 1593, again suggesting that this practice was common then. But most of the historical references to homosexuality relate either to great men or to great scandals. However, over the last seventy years or so various scholars have tried to trace the history of sex,[18] and it is possible to glean a good deal from what they have found and also from what they have failed to establish.

Their studies of English history before the seventeenth century consist usually of inconclusive speculation as to whether certain men, such as Edward II, Christopher Marlowe, William Shakespeare, were or were not homosexual. Yet the disputes are inconclusive not because of lack of evidence but because none of these men fits the modern stereotype of the homosexual.

It is not until the end of the seventeenth century that other kinds of information become available and it is possible to move from speculations about individuals to descriptions of homosexual life. At this period references to homosexuals as a type and to a rudimentary homosexual subculture, mainly in London, begin to appear. But the earliest descriptions of homosexuals do not coincide exactly with the modern conception. There is much more stress on effeminacy and in particular in transvestism, to such an extent that there seems to be no distinction at first between transvestism and homosexuality.[19] The terms emerging

---

[15] Ford and Beach, *op. cit.*, p. 132.

[16] *Ibid.*, pp. 131-132.

[17] Geoffrey May, *Social Control of Sex Expression*, London: Allen and Unwin, 1930, pp. 65 and 101.

[18] Especially Havelock Ellis, *Sexual Inversion*, London: Wilson and Macmillan, 1897; Iwan Bloch (E. Dühren, pseud.), *Sexual Life in England Past and Present*, English translation, London: Francis Aldor, 1938; German edition, Charlottenberg, Berlin, 1901-03; Gordon Rattray Taylor, *Sex in History*, London: Thames and Hudson, 1953; Noel I. Garde, *Jonathan to Gide: The Homosexual in History*, New York: Vantage, 1964.

[19] Dr. Evelyn Hooker has suggested that in a period when homosexual grouping and a homosexual subculture have not yet become institutionalized, homosexuals are likely to behave in a more distinctive and conspicuous manner because other means of making contact are not available. This is confirmed by the fact that lesbians are more

at this period to describe homosexuals —Molly, Nancy-boy, Madge-cull—emphasize effeminacy. In contrast the modern terms—like fag, queer, gay, bent —do not have this implication.[20]

By the end of the seventeenth century, homosexual transvestites were a distinct enough group to be able to form their own clubs in London.[21] Edward Ward's *History of the London Clubs*, published in 1709, describes one called "The Mollies' Club" which met "in a certain tavern in the City" for "parties and regular gatherings." The members "adopt(ed) all the small vanities natural to the feminine sex to such an extent that they try to speak, walk, chatter, shriek and scold as women do, aping them as well in other respects." The other respects apparently included the enactment of marriages and child-birth. The club was discovered and broken up by agents of the Reform Society.[22] There were a number of similar scandals during the course of the eighteenth century as various homosexual coteries were exposed.

A writer in 1729 descibes the widespread homosexual life of the period:

> They also have their Walks and Appointments, to meet and pick up one another, and their particular Houses of Resort to go to, because they dare not trust themselves in an open Tavern. About twenty of these sort of Houses have been discovered, besides the Nocturnal Assemblies of great numbers of the like vile Persons, what they call the *Markets*, which are the Royal Exchange, Lincoln's Inn, Bog Houses, the south side of St. James's Park, the Piazzas in Covent Garden, St. Clement's Churchyard, etc.
>
> It would be a pretty scene to behold them in their clubs and cabals, how they assume the air and affect the name of Madam or Miss, Betty or Molly, with a chuck under the chin, and "Oh, you bold pullet, I'll break your eggs," and then frisk and walk away.[23]

The notion of exclusive homosexuality became well-established during this period. When "two Englishmen, Leith and Drew, were accused of paederasty . . . .The evidence given by the plaintiffs was, as was generally the case in these trials, very imperfect. On the other hand the defendants denied the accusation, and produced witnesses to prove their predelection for women. They were in consequence acquitted."[24] This could only have been an effective argument in a society that perceived homosexual behavior as incompatible with heterosexual tastes.

During the nineteenth century there are further reports of raided clubs and homosexual brothels. However, by this time the element of transvestism had diminished in importance. Even the male prostitutes are described as being of masculine build and there is more stress upon sexual license and less upon dressing up and play-acting.

The Homosexual Role and Homosexual Behavior

Thus, a distinct, separate, specialized role of "homosexual" emerged in England at the end of the seventeenth century and the conception of homosex-

---

conspicuous than male homosexuals in our society, but does not seem to fit the 17th century, where the groups are already described as "clubs."

[20] However, "fairy" and "pansy," the commonest slang terms used by non-homosexuals, have the same meaning of effeminate as the earlier terms.

[21] Bloch, *op. cit.*, p. 328, gives several examples, but attributes their emergence to the fact that "the number of homosexuals increased."

[22] Quoted in *ibid.*, pp. 328-329.

[23] Anon, *Hell upon Earth: or the Town in an Uproar*, London, 1729, quoted by G. R. Taylor in Marmor, editor, *op. cit.*, p. 142.

[24] Bloch, *op. cit.*, p. 334.

uality as a condition which characterizes certain individuals and not others is now firmly established in our society. The term role is, of course, a form of shorthand. It refers not only to a cultural conception or set of ideas but also to a complex of institutional arrangements which depend upon and reinforce these ideas. These arrangements include all the forms of heterosexual activity, courtship, and marriage as well as the labeling processes—gossip, ridicule, psychiatric diagnosis, criminal conviction—and the groups and networks of the homosexual subculture. For simplicity we shall simply say that a specialized role exists.

How does the existence of this social role affect actual behavior? And, in particular, does the behavior of individuals conform to the cultural conception in the sense that most people are either exclusively heterosexual or exclusively homosexual? It is difficult to answer these questions on the basis of available evidence because so many researchers have worked with the preconception that homosexuality is a condition, so that in order to study the behavior they have first found a group of people who could be identified as "homosexuals." Homosexual behavior should be studied independently of social roles, if the connection between the two is to be revealed.

This may not sound like a particularly novel program to those who are familiar with Kinsey's contribution to the field.[25] He, after all, set out to study "sexual behavior;" he rejected the assumptions of scientists and laymen:

that there are persons who are "heterosexual" and persons who are "homosexual", that these two types represent antitheses in the sexual world and that there is only an insignificant class of "bisexuals" who occupy an intermediate position between the other groups . . . that every individual is innately—inherently—either heterosexual or homosexual . . . (and) that from the time of birth one is fated to be one thing or the other . . . .[26]

But, although some of Kinsey's ideas are often referred to, particularly in polemical writings, surprisingly little use has been made of his actual data.

Most of Kinsey's chapter on the "Homosexual Outlet"[27] centers on his "heterosexual-homosexual rating scale." His subjects were rated on this scale according to the proportion of their "psychologic reactions and overt experience" that was homosexual in any given period of their lives. It is interesting, and unfortunate for our purposes, that this is one of the few places in the book where Kinsey abandons his behavioristic approach to some extent. However, "psychologic reactions" may well be expected to be affected by the existence of a social role in the same way as overt behavior. Another problem with using Kinsey's material is that although he gives very full information about sexual behavior, the other characteristics of the people he interviewed are only given in a very bald form.[28] But Kinsey's study is un-

---

[25] Alfred C. Kinsey et al., *Sexual Behavior in the Human Male*, Philadelphia and London: Saunders, 1948; and Kinsey et al., *Sexual Behavior in the Human Female*, Philadelphia and London: Saunders, 1953.

[26] Kinsey et al., *Sexual Behavior in the Human Male*, pp. 636-37.

[27] *Ibid.*, ch. 21, pp. 610-666.

[28] The more general drawbacks of Kinsey's data, particularly the problem of the representativeness of his sample, have been thoroughly canvassed in a number of places; see especially William G. Cochran et al., *Statistical Problems of the Kinsey Report on Sexual Behavior in the Human Male*, Washington: American Statistical Society, 1954.

doubtedly the fullest description there is of sexual behavior in any society and as such it is the safest basis for generalizations to other Western societies.

The ideal way to trace the effects on behavior of the existence of a homosexual role would be to compare societies in which the role exists with societies in which it does not. But as there are no adequate descriptions of homosexual behavior in societies where there is no homosexual role, we shall have to substitute comparisons within American society.

(1) *Polarization*

If the existence of a social role were reflected in people's behavior, we should expect to find that relatively few people would engage in bisexual behavior. The problem about investigating this empirically is to know what is meant by "relatively few." The categories of Kinsey's rating scale are, of course, completely arbitrary. He has five bisexual categories, but he might just as well have had more or less, in which case the number falling into each would have been smaller or larger. The fact that the distribution of his scale is U-shaped, then, is in itself meaningless. (See Table 1).

It is impossible to get direct evidence of a polarization between the homosexual and the heterosexual pattern, though we may note the suggestive evidence to the contrary that at every age far more men have bisexual than exclusively homosexual patterns. However, by making comparisons between one age group and another and between men and women, it should be possible to see some of the effects of the role.

(2) *Age Comparison*

As they grow older, more and more men take up exclusively heterosexual patterns, as Table 1, Column 2 shows. The table also shows that *each* of the bisexual and homosexual categories, columns 3-8, contains fewer men as time goes by after the age of 20. The greatest losses are from the fifth bisexual category, column 7, with responses that are "almost entirely homosexual." It is a fairly small group to

TABLE 1
HETEROSEXUAL-HOMOSEXUAL RATING: ACTIVE INCIDENCE BY AGE*

	Percent of each age group of male population having each rating								
	(1)	(2)	(3)	(4)	(5)	(6)	(7)	(8)	(9)
Age	X	0	1	2	3	4	5	6	1-6
15	23.6	48.4	3.6	6.0	4.7	3.7	2.6	7.4	28.0
20	3.3	69.3	4.4	7.4	4.4	2.9	3.4	4.9	27.4
25	1.0	79.2	3.9	5.1	3.2	2.4	2.3	2.9	19.8
30	0.5	83.1	4.0	3.4	2.1	3.0	1.3	2.6	16.4
35	0.4	86.7	2.4	3.4	1.9	1.7	0.9	2.6	12.9
40	1.3	86.8	3.0	3.6	2.0	0.7	0.3	2.3	11.9
45	2.7	88.8	2.3	2.0	1.3	0.9	0.2	1.8	8.5

* Based on Kinsey (1948) p. 652, Table 148.
X = unresponsive to either sex; 0 = entirely heterosexual; 1 = largely heterosexual, but with incidental homosexual history; 2 = largely heterosexual but with a distinct homosexual history; 3 = equally heterosexual and homosexual; 4 = largely homosexual but with distinct heterosexual history; 5 = largely homosexual but with incidental heterosexual history; 6 = entirely homosexual.

begin with, but by the age of 45 it has almost entirely disappeared. On the other hand the first bisexual category, column 3, with only "incidental homosexual histories" has its numbers not even halved by the age of 45. Yet at all ages the first bisexual category represents a much smaller proportion of those who are almost entirely homosexual (columns 2 and 3) than the fifth category represents of those who are almost entirely homosexual (columns 7 and 8). In everyday language, it seems that proportionately more "homosexuals" dabble in heterosexual activity than "heterosexuals" dabble in homosexual activity and such dabbling is particularly common in the younger age groups of 20 to 30. This indicates that the existence of the despised role operates at all ages to inhibit people from engaging in occasional homosexual behavior, but does not have the effect of making the behavior of many "homosexuals" exclusively homosexual.

On the other hand, the overall reduction in the amount of homosexual behavior with age can be attributed in part to the fact that more and more men become married. While the active incidence of homosexual behavior is high and increases with age among single men, among married men it is low and decreases only slightly with age. Unfortunately the Kinsey figures do not enable us to compare the incidence of homosexuality among single men who later marry and those who do not.

(3) Comparison of Men and Women

The notion of a separate homosexual role is much less well-developed for women than it is for men and so too are the attendant techniques of social control and the deviant subculture and organization. So a comparison with women's sexual behavior should tell us something about the effects of the social role on men's behavior.

Fewer women than men engage in homosexual behavior. By the time they are 45, 26 percent of women have had *some* homosexual experience, whereas about 50 percent of men have. But this is probably a cause rather than an effect of the difference in the extent to which the homosexual role is crystallized, for women engage in less non-marital sexual activity of any kind than men. For instance, by the time they marry 50 percent of women have had some pre-marital heterosexual experience to orgasm, whereas as many as 90 percent of men have.

The most revealing contrast is between the male and female distributions on the Kinsey rating scale, shown in Table 2. The distributions for women follow a smooth J-shaped pattern, while those for men are uneven with an increase in numbers at the exclusively homosexual end. The distributions for women are the shape that one would expect on the assumption that homosexual and heterosexual acts are randomly distributed in a ratio of 1 to 18.[29] The men are relatively more concentrated in the exclusively homosexual category. This appears to confirm the hypothesis that the existence of the role is reflected in behavior.

Finally, it is interesting to notice that although at the age of 20 far more men than women have homosexual and bisexual patterns (27 percent as against 11 percent), by the age of 35 the figures are both the same (13 percent). Women seem to broaden their sexual experience as they get older

---

[29] This cannot be taken in a rigorously statistical sense, since the categories are arbitrary and do not refer to numbers, or even proportions, of actual sexual acts.

TABLE 2
COMPARISON OF MALE AND FEMALE HETEROSEXUAL-HOMOSEXUAL RATINGS: ACTIVE INCIDENCE AT SELECTED AGES*

	Age	Percent of each age group having each rating								
		(1) X	(2) 0	(3) 1	(4) 2	(5) 3	(6) 4	(7) 5	(8) 6	(9) 1-6
Male	20	3.3	69.3	4.4	7.4	4.4	2.9	3.4	4.9	27.4
Female		15	74	5	2	1	1	1	1	11
Male	35	0.4	86.7	2.4	3.4	1.9	1.7	0.9	2.6	12.9
Female		7	80	7	2	1	1	1	1	13

* Based on Kinsey (1948) p. 652, Table 148 and Kinsey (1953) p. 499, Table 142. For explanation of the ratings, see Table 1.

whereas more men become narrower and more specialized.

None of this, however, should obscure the fact that, in terms of behavior, the polarization between the heterosexual man and the homosexual man is far from complete in our society. Some polarization does seem to have occurred, but many men manage to follow patterns of sexual behavior that are between the two, in spite of our cultural preconceptions and institutional arrangements.

CONCLUSION

This paper has dealt with only one small aspect of the sociology of homosexuality. It is, nevertheless, a fundamental one. For it is not until he sees homosexuals as a social category, rather than a medical or psychiatric one, that the sociologist can begin to ask the right questions about the specific content of the homosexual role and about the organization and functions of homosexual groups.[30] All that has been done here is to indicate that the role does not exist in many societies, that it only emerged in England towards the end of the seventeenth century, and that, although the existence of the role in modern America appears to have some effect on the distribution of homosexual behavior, such behavior is far from being monopolized by persons who play the role of homosexual.

[30] But an interesting beginning has been made by Evelyn Hooker in "The Homosexual Community," *Proc. XIVth Int. Congr. Appl. Psychol. Personality Research*, Vol. 2, Copenhagen, Munksgaard, 1962; and "Male Homosexuals and their Worlds," Marmor, editor, *op. cit.*, pp. 83-107; there is much valuable descriptive material in Donald Webster Cory, *The Homosexual in America*, New York: Greenberg, 1951; and in Gordon Westwood, *A Minority: A Report on the Life of the Male Homosexual in Great Britain*, London: Longmans, 1960, as well as elsewhere.

Chapter Five

Gay Liberation and Lesbian Feminism

From the Stonewall Rebellion . . .

> Liberation for gay people is to define for ourselves how and with whom we live, instead of measuring our relationships by straight values. . . . To be a free territory, we must govern ourselves, set up our own institutions, defend ourselves, and use our own energies to improve our lives.
>
> —Carl Wittman, *Refugees from Amerika: A Gay Manifesto*

On the Friday night of 27–28 June 1969, New York police raided a Greenwich Village gay bar called the Stonewall. Bar raids were an American institution—a police rite to "manage" the powerless and disrespectable—and in the preceding three weeks, five New York gay bars had already been raided. What made the Stonewall a symbol of a new era of gay politics was the reaction of the drag queens, dykes, street people, and bar boys who confronted the police first with jeers and high camp and then with a hail of coins, paving stones, and parking meters. By the end of the weekend, the Stonewall bar had been burned out, but a new form of collective resistance was afoot: gay liberation. The Mattachine Action Committee responded to the Stonewall outbreak with a flier on 29 June calling for organized resistance, and within a few days radical students at the Alternative University were providing meeting space for a Gay Liberation Front (Teal 1971, 17–23; Marotta 1981, 72–85).

Still, Stonewall was no isolated event. A police campaign against Los Angeles gay bars in 1967 had sparked a rally of several hundred "on Sunset Boulevard, where they listened to angry speakers intoning the phrases of confrontational politics" (D'Emilio 1983, 227), and student activism, especially on the campuses of Columbia University and the Sorbonne, were associated with the formation of radical gay caucuses. In 1967 and 1968, political tensions were mounting to new heights with clashes between police and black nationalists, hippies, students, and antiwar demonstrators, most notably at the National Democratic Convention in Chicago. In the Netherlands, the Socialist Youth formed a gay caucus, and student groups openly sponsored gay dances on campus (Straver 1973, 170–72). Student Homophile Leagues were formed in 1967 at Columbia by Robert A. Martin and at New York University by Rita Mae Brown. In 1968, the Columbia group picketed a psychiatric seminar on homosexuality held on campus (Martin 1983). In May of the same year, Paris erupted in a general strike and students seized the campus of the Sorbonne in a protest that shared New Left goals. Amidst the "liberated zones," a Comité d'Action Pédératique Révolutionnaire met, much to the dismay of the orthodox Left (Girard 1981, 80).

The new militants, then, typically came out of student and other New Left movements and carried with them current debates and precepts, in which they turned to issues of gender and sexuality. Radicalized by their experiences in black and student organizations, they were now thinking through their own lives with new concepts and were taking a militant message to new constituencies. Feminists and gay liberationists often thought of themselves as revolutionaries rejecting a fundamentally unequal and corrupt power establishment in favor of participatory democracy whereby all the voiceless and suppressed could gain a measure of control over their own lives. Civil rights had become passé: why petition to be let into a social system so deeply riven by racism, sexism, militarism, and heterosexism?

The goal that radical women and gay men shared with the counterculture was "to construct community institutions based on democratic participation": free universities, an underground press, communes, a society of cooperative and nonexploitative relations (see Breines 1982). Deeply suspicious of leaders, bureaucracies, and political parties, the fundamental movement unit was the consciousness-raising group. As explained in

Come Out!, the journal of the New York Gay Liberation Front, it was a deceptively simple mechanism:

A consciousness raising group is a group of gay people who have regular sessions together. By consensus a topic is selected for each session. Each member of the group contributes her personal experiences relating to the chosen topic. When all of the testimony is heard, the group locks into the similarity in the experiences related by all the members. . . . A gay person begins to see that his personal hang-ups, those that he was afraid to divulge to others, are indeed the same hang-ups that other gays were also afraid to divulge. It becomes increasingly difficult to explain this commoness without considering each person's interactions with sexist society. (Gavin 1971, 19)

The group's chairperson would be selected by lot and rotated from meeting to meeting. To limit the formation of elites, every person in the group would be given the floor in turn. Analysis of one's own situation was to flow from the collective experience, owing nothing to received dogmas. Consciousness raising was a technique well known from the "speaking bitterness" campaigns of the Chinese cultural revolution, and, as in China, it helped empower the powerless and grant participation to the masses.[1]

The result of these intense discussions was immense anger, joy, pride, and a boiling over of new ideas. People glimpsed the future and fell in love with a utopia far from the bad old days with their repression and terror, hiding and fear. Gay liberation groups rarely reached the consensus they assumed would come out of consciousness raising, but stimulated outpourings of hopes and ambitions of irreconcilable diversity. Resolutely guarding itself against stasis, gay liberation in its heyday—from 1969 to 1972—functioned as an ongoing catalyst. Like the New Left itself, which had spawned new social movements, gay liberation ultimately was to produce a larger set of gay and lesbian groups.

Sexuality was a yet undeveloped theme in radical thought. In addition to the Beat poets, New Left figures such as Paul Goodman and Daniel Cohn-Bendit had raised it at various times as did a few relatively isolated European intellectuals. Simone de Beauvoir's *Second Sex* raised many of the issues of modern feminism two decades before the revival of the modern movement, and Herbert Marcuse, who had been a youthful participant in the 1918 German revolution and had been steeped in the thinking of the life-reform movements of the Weimar Republic, caught the imagination of many gay liberationists. His *Eros and Civilization*, published in the ideological wasteland of 1955, bridged the prewar and postwar gay movements with its implicit vision of homosexuality as a protest "against the repressive order of procreative sexuality" and as an affirmation of a liberated sensualism (37, 155, 183). As well, Allen Ginsberg, testifying at the trial of black and student movement leaders arrested at the Chicago Democratic convention, invoked the socialist fraternalism of Whitman and Carpenter. In the face of the prosecutor's characterization of the Chicago protestors as "freaking fag revolutionaries," Ginsberg spoke out for

a natural tenderness between all citizens, not only men and women but also a tenderness between men and men as part of our democratic heritage, part of the Adhesiveness which would nake the democracy function: that men could work together not as competitive beasts but as tender lovers and fellows. (Tytell 1976, 243; Ginsberg 1974, 14)

Gay liberation never thought of itself as a civil rights movement for a particular minority but as a revolutionary struggle to free the homosexuality in everyone, challenging the conventional arrangements that confined sexuality to heterosexual monogamous families. For gay liberation there was no "normal" or "perverse" sexuality, only a world of sexual possibilities ranged against a repressive order of marriage, oedipal families, and compulsory heterosexuality. It is in this context that Dennis Altman could foresee an "end of the homosexual" because "gay liberation will succeed as its raison d'être disappears" (Altman 1971, 225; see Front Homosexual 1971). Once everyone was free to express her or his latent sexualities, boundaries between the homosexual and the heterosexual should fade into irrelevance and false partitions in the flow of desire give way to personal fulfillment.

Carl Wittman's 1970 "Gay Manifesto" drew together many of the themes of gay liberation thinking. Announcing "we are euphoric, high, with the initial flourish of a movement," it began, "we have to realize that our loving each other is a good thing." Characterizing San Francisco as a "refugee camp" and a "ghetto" controlled by the heterosexist occupational forces of law, police, employers, and capital, Wittman called for rejection of heterosexual standards of gender and monogamy, an end to homophile conformity and closetry, resistance to street violence and police harassment, and confrontation with the "psychological warfare" purveyed by the mass media. "We strive," he continued, "for democratic,

mutual, reciprocal sex," affirming the possibility of this ideal even in manboy and sadomasochistic relationships. Gay liberation also meant coalition with other progressive forces, especially feminism, as well as with black, Chicano, radical, hip, and homophile movements (1972, 157–71).

Gay liberation groups sprang up in the spring and summer of 1969 in the San Francisco Bay area and New York City. Leo Laurence forwarded the radical plank in the pages of SIR's *Vector* and came out with his lover in the countercultural *Berkeley Barb*. The upshot was his lover's dismissal from his job with a steamship company and Laurence's removal from the editorship of *Vector*. They then formed a Committee for Homosexual Freedom, which picketed the steamship company and then a record store that had also fired a gay employee. The Stonewall Rebellion in New York engendered a wave of new groups willing to take immediate direct action against the old array of antihomosexual institutions. In late summer, the New York Gay Liberation Front (NYGLF) and the Mattachine Action Committee picketed in a park where trees had been cut down to eliminate cruising (that is, gay men meeting each other). The GLF joined in antiwar rallies and presented the new platform to the 1969 North America Conference of Homophile Organizations (NACHO) in Kansas City. By fall, GLF dances were regular events in New York, Chicago, and Berkeley, cities where men had often been arrested for dancing or touching in public. Pickets arrived at the *Village Voice* protesting its refusal to print the word *gay* and at *Time* magazine and the *San Francisco Examiner* for their demeaning treatment of gay people. Newspapers such as *Gay Power*, *Come Out!*, and *Gay* sprang out of movement committees. The GLF confronted Western and Delta airlines about their employment practices, and SIR picketed Macy's for having gay men entrapped by police in its washrooms. Transvestites formed Street Transvestite Action Revolutionaries, and blacks and Hispanics organized Third World Gay Revolution. At the end of the first year, two to three thousand marched to Central Park in New York to commemorate the Stonewall Rebellion as did hundreds in Los Angeles and Chicago (see Teal 1971; D'Emilio 1983; Humphreys 1972b).

In 1970, after "three terrible, joyous days of open, honest battle," conflicts between gay liberation and the old guard wrenched apart a NACHO meeting in San Francisco. In the end, the conference "passed motions supporting women's liberation and the Black Panthers, calling for immediate withdrawal of American forces from Vietnam, authorizing a Gay Strike Day, and calling for memorialization of homosexuals killed in Nazi concentration camps" (Rankin 1970, 4; Humphreys 1972b, 108).

In the same month, the Black Panther leader, Huey Newton, declared his solidarity for the gay movement, stating that "homosexuals are not given freedom and liberty by anyone in the society. Maybe they might be the most oppressed people in the society" (1972, 195). Gay and lesbian delegates, in turn, showed up at the Panther-sponsored Revolutionary Peoples' Constitutional Convention in September to claim their place in the radical coalition that so upset the ruling elites of the United States. But as early as November 1969, GLF experienced a schism. Jim Owles and Marty Robinson walked out to found the Gay Activists Alliance (GAA) in New York, having found the GLF too anarchic and self-focused, strong on rhetoric but unable to plan effectively, and too preoccupied with revolutionary doctrine to address the day-to-day discrimination occurring around it. The GAA wanted to concentrate on the one issue of gay rights without the diffusion of energy into other New Left causes evident in the GLF. For the GLF, the GAA represented a regression to homophile accommodationism and an abandonment of total social transformation for piecemeal reform. The GAA's adoption of a committee structure and elected leadership, they believed, betrayed the GLF's commitment to consensus and participatory democracy. The movement was facing a transition experienced by so many others before it, when charisma and chiliasm give way to structure and institution. In the end, the GAA proved more durable and effective and the GLF soon exhausted itself (see Altman 1971, 116; Humphreys 1972b, 124; Teal 1971, 106; Marotta 1981, 150).

In practice, many participants flowed between both organizations and the two cooperated on a number of projects. Renewed bar raids in March 1970 brought another round of street demonstrations. Election candidates faced sharp questions on gay rights, and GAA activists forced the New York mayor to address gay issues before television and opera audiences. City hall, the *New York Post*, *Harper's*, the *New York Times*, and the "Dick Cavett Show" felt the wrath of the GAA in 1970 and 1971 "zaps," or confrontations. The GAA set up task-oriented committees on political action, police, elections, civil rights law for the city, fair taxes, law, news, leaflets and graphics, fund-raising, social affairs, and member orientation—soon accumulating a thick dossier on antihomosexual discrimination.

The GAA's response was often ingenious: "In the summer of 1971, the owner of a credit agency on New York's 42nd Street was questioned about his agency's practice of informing employers of the suspected homosexual tendencies of prospective employees, as well as credit applicants. When questioned about how he determined sexual orientation, he was quoted as saying, "'If a man looks like a duck, walks like a duck, quacks like a duck, and associates with ducks, I'd say he is a duck.' In a short time, a dozen GAA members dressed in duck costumes were waddling around the sidewalk at the entrance to the credit agency, quacking and carrying picket signs" (Humphreys 1972b, 126).

Perhaps the best-known success of the early 1970s was the assault mounted against American psychiatry, which resulted in the 1973–74 removal of homosexuality from the American Psychiatric Association's official diagnostic manual. A century of psychiatric talk in the United States had provided the underpinnings for a range of anti-homosexual practices. After all, what rights could a psychopathology have? If gay men and lesbians were no more than diseased beings, then state institutions had a duty to stamp them out by isolating them in prisons and hospitals, excluding them from a wide range of employment, barring them from entering the country, banning them from bars, and suppressing their voices in the arts and literature. *One* magazine had long disdained psychiatric ideology, but it was not until the militant 1970s that gay people gained sufficient strength and confidence to confront the therapeutic establishment directly. In 1968, even before Stonewall, a contingent of San Franciscans arrived unannounced at a convention of the American Medical Association to speak out against the scientific extermination of homosexuality. In the same year, students demanded of a medical forum at Columbia University that "it is time that talk stopped being *about* us and started being *with* us" (see Teal 1971, 293–97; Kameny 1969; Bayer 1981, 92). Gay liberation fronts stormed San Francisco, Los Angeles, and Chicago conventions of psychiatry, medicine, and behavior modification in 1970, where sessions on the "treatment" and "correction" of homosexuality were disrupted with cries of "barbarism," "medieval torture," and "disgusting" and with demands for equal time.

These GLF zaps rapidly polarized the psychiatric profession between such hard-line conservatives as Edward Bergler, Irving Bieber, Charles Socarides, Lionel Ovesey, and Lawrence Hatterer (whom Allen Young characterized as the "war criminals") and a growing liberal contingency including Ernest Van Den Haag, Hendrik Ruitenbeek, and George Weinberg, who had been questioning the psychiatric label for some years. An unprecedented panel of gay people was arranged for the 1971 convention of the APA in Washington, D.C., where Frank Kameny, Larry Littlejohn of SIR, Del Martin of DOB, Lille Vincenz, and Jack Baker, president of the University of Minnesota Students' Association, represented the movement. A 1972 panel included liberal psychiatrists and a gay psychiatrist who appeared wearing a mask. The issue reached a climax in 1973 with a debate between Irving Bieber and Charles Socarides on one side and Judd Marmor, Richard Green, Robert Stoller, and Ron Gold on the other. Gold's paper, "Stop! You're Making Me Sick," represented the gay movement's position.

Official changes were already underway elsewhere as the American Sociological Association passed a no-discrimination resolution in 1969; the National Association for Mental Health called for decriminalization in 1970; the states of Connecticut, Colorado, and Oregon did decriminalize in 1971; a federal court stopped automatic dismissal of gay people from federal employment in the same year; and the National Association of Social Workers rejected the medical model of homosexuality in a 1972 resolution. As the GAA waned through internal dissension (coming to an end, at least symbolically, when its community center was fire-bombed in 1974), leading movement activists reorganized as the National Gay Task Force to press forward the antipsychiatric struggle. When the APA Council accepted deletion of homosexuality from the diagnostic manual in a unanimous vote in 1973, the conservatives forced a referendum on the issue. The result of this curious spectacle of defining pathology by plebiscite was a vote of 58 percent for deletion and 37 percent for retention in 1974. In the end, the new diagnostic manual included a compromise category that continued to allow psychiatrists to "treat" people unhappy with their sexual orientation.

The movement forced debate on homosexuality among a number of professional and scholarly associations in the 1970s, opening the way for the formation of gay and lesbian caucuses within several disciplines: librarianship in 1970; modern languages and psychology in 1973; sociology in 1974; history, psychiatry, and public health in 1975; nursing and social work 1976; and a general Gay Academic Union in 1973 (see Noll 1978, 173–77).

. . . to a World Movement

Within two years from the Stonewall Rebellion, gay liberation groups emerged in every major city and campus in the United States, Canada, Australia, and Western Europe. With a gay liberation press founded in Los Angeles (*Advocate*), New York (*Come Out!*), San Francisco (*Gay Sunshine*), Boston (*Fag Rag*), Detroit (*Gay Liberator*), Toronto (*Body Politic*), and London (*Come Together*), far-flung organizations became much more connected and aware of diverse initiatives. On three conti-

nents, gay movements in the early 1970s developed along a similar course, with parallel Left-oriented gay liberation groups forming along with more liberal civil rights organizations. With the general decline of New Left movements in the late 1970s, self-professed gay liberation fronts faded as well, leaving reformist groups in the political field and engendering a new proliferation of gay and lesbian interest groups organized within existing institutions: in the workplace, church, the theater, social services, business, and sports.

The British experience illustrates the process in the early 1970s. The North-Western Committee of the Homosexual Law Reform Society reconstituted itself as the Committee (and then, Campaign) for Homosexual Equality (CHE) in 1969, adopting a platform aimed "to remove fear, discrimination and prejudice against homosexuals, to achieve full equality before the law, and to promote the positive acceptance of homosexuality as a valid way of life" (Marshall 1980, 78). A successful, nonthreatening formula, it attracted sixty local groups by 1972, which offered telephone counseling, regular discos and meeting places, and a concrete political agenda: equalization of the age of consent at sixteen, extension of the 1967 decriminalization to the military, to Scotland, and to Ulster; abolition of gross indecency laws, and freedom of the gay press (see Weeks 1977, 207–13; Galloway 1983).

Gay liberation arrived in London in 1970, when Aubrey Walter and Bob Mellors returned from New York to call a gay liberation meeting at the London School of Economics. Like its American counterparts, the London GLF evolved through high-energy consciousness-raising groups into a collection of workshops focusing on antihomosexual practices in psychiatry, the church, and government. Soon it was working on public education, women's and youth issues, the media, and street theater. Coming out, or public confrontation of its antagonists was always a central feature of gay liberation. As well as forcing its persecutors to become aware of the maliciousness of their actions, coming out had an immensely exhilarating and self-healing effect upon gay men and lesbians who had, for so long, lived a secretive and shamed existence (see Adam 1978, 126). Essential for personal and social change was gay pride, asserting the worth and capability of a people rejected as despicable and weak. For the London GLF, its first act of coming out took the form of a November 1970 demonstration in "Highbury Fields," where a prominent Young Liberal had been arrested by the police and accused of 'indecency'" (Walter 1980, 12). At its height in 1971, the GLF was active in Birmingham, Manchester, Bristol, Cardiff, Edinburgh, and Leeds. But the GLF was all but defunct by the end of 1972, torn apart by tensions between women and men, drag queens and machos, socialists and counterculturalists. By drawing together such a diversity of gay people and engendering such utopian aspirations, the GLF could not resolve the intensely different experiences of its adherents. If male domination was the problem as the feminists and effeminists agreed, then rejection of masculinity was the solution and many GLF men briefly embraced "gender-fuck" drag—mixing beards and dresses, jewelry and leather—in order to parody gender. If sexual repression and the nuclear family were the problem, then public affection and sexual communism could be the answer. In the end, few could so radically rearrange their emotional lives, and such experiments proved more dramatic than viable. At the personal level, many who had come out for companionship and community experienced too much hostility and pain in the GLF cauldron to want to continue devoting so much of themselves to the cause, and GLF yielded to CHE's more sober and limited style.

In Canada, sporadic homophile groups had come about as early as 1964 with the Vancouver Association for Social Knowledge and in 1965 with the Ottawa Council on Religion and the Homosexual. A group of six, who wrote an open letter to Toronto newspapers and to Liberal prime minister Lester Pearson, opened the question of decriminalization in 1964. But it was not until 1967, when the British Parliament approved a new Sexual Offenses Act, that debate in Canada began in earnest. Also in 1967, the Supreme Court upheld the indefinite sentence of a Northwest Territories man, Everett Klippert, as a "dangerous sexual delinquent" following repeated convictions for consenting sexual relations with adult men. The following year, the justice minister, Pierre Trudeau, promised law reform, stating that "the state has no place in the bedrooms of the nation" (Sylvestre 1979, 24), and in August 1969 a new "consenting adults in private" law was proclaimed following passage by the Liberal and New Democratic parties in Parliament. (Many Conservatives and the right-wing Parti Créditiste voted no.)

The modern gay and lesbian movement took the familiar route. Campus groups organized first at the University of Toronto in 1969 and, within three years, across the nation. A Gay Liberation Front formed in 1970 in the well-developed counterculture of Vancouver and then in 1972 in Montreal (Front de Libération Homosexuel) and Toronto (Gay Action). In 1971, a group around George Hislop staked out more moderate ground with the Community Homophile Association of Toronto (CHAT).

When the first march on Parliament was held in 1971, Gays of Ottawa enunciated its law reform program: abolition of the gross indecency law, a uniform age of consent, protection through the human rights codes, equal rights for homosexual couples, destruction of police files, and the ending of discrimination in immigration, employment, custody and adoption, and housing (Jackson and Persky 1982, 217–20). A national meeting in 1972 to plan strategy for a federal election led to annual meetings coordinated by a National Gay Rights Coalition.

As in the United States and the United Kingdom, gay organizations unfolded in Canada throughout the 1970s even in small towns and rural areas where, for the first time, they often *preceded* the commercial infrastructure of bars and public meeting places. In small cities, such as Saskatoon (in 1973) and London, Ontario (in 1974), community-run clubhouses offered the first gay and lesbian places in their regions (see Warner 1976). In sparsely populated areas, such as Newfoundland, northern Ontario, and the British Columbia interior, the urban press provided the catalyst to overcome geography and connect widely dispersed gay and lesbian readers.

In Australia and New Zealand, gay and lesbian organization showed much the same pattern of development as its kin in the rest of the English-language world. With the deepening involvement of Australian forces in the war against Vietnam, an antiwar movement mobilized through the late 1960s, opening an intense political debate and a crisis of confidence in the entrenched Liberal administration. Homophile groups surfaced briefly in 1969 with a chapter of the Daughters of Bilitis in Melbourne and an Australian Capital Territory Homosexual Law Reform Society. A more enduring homophile group, the Campaign Against Moral Persecution (CAMP) formed in Sydney through the initiative of John Ware and Christabell Poll, which stressed the "ordinariness of homosexuality" and sought reform through public education (see Thompson 1985, 10; Johnston 1984; Altman 1979). The group quickly formed chapters in the other state capitals, issuing a journal, *CAMP Ink*, from 1971. In its first demonstration in October 1971, CAMP targeted Liberal party headquarters in Sydney to challenge the preselection candidacy of an opponent of homosexual law reform. When an election was called in 1972, a gay activist ran against the Liberal prime minister, garnering 218 votes. (A similar attempt was made against the New Zealand prime minister to publicize gay concerns.) The ensuing Labour party government decriminalized homosexuality the following year in areas of federal jurisdiction, the Australian Capital and Northern Territories. With gay liberation splitting from CAMP in the mid-seventies, public actions against media, church, and government reached a height only to die down by 1975–76.

The postwar hegemony of the United States, especially among the advanced capitalist nations, as well as among much of the third world, has also had an impact upon the social organization of homosexuality and the development of a political movement. But national traditions and varying arrays of social preconditions have led to different paths of movement development. As argued earlier, a complex set of socioeconomic factors and political possibilities created the crucible in which homosexuality became organized into gay and lesbian subcultures in Western countries. With a shared language, cultural diffusion became an important stimulus for parallel development of the gay world and its movement in the United States, United Kingdom, Canada, Australia, and New Zealand. (Despite its nationalism, Quebec cannot help but be deeply influenced by the Anglo-American culture that surrounds it.) Among other language communities and among nations with different political legacies and economic systems, the movement, although cognizant of the Stonewall heritage, has developed along alternative paths.

In the Federal Republic of Germany, student activism and the coming to power of the Social Democrats preceded the emergence of the modern feminist and gay movements. With the end of the cold war Christian Democrat government, the Social Democrats decriminalized homosexuality in 1969, later lowering the age of consent from twenty-one to eighteen in 1973. Campus action groups (*Aktionsgruppen*) sprang up across the country in 1971–73, often following screenings of Rosa von Praunheim's controversial film, *Not the Homosexual Is Perverse, But the Situation in Which He Lives*, which documented the gay upheaval in the United States. Among the first was Homosexuelle Aktion Westberlin, which adopted an explicitly radical approach (see Stümke and Finkler 1981, 410–14; Dannecker and Reiche 1973).

In the Netherlands, a peculiar balance of political forces that has guaranteed a more genuinely pluralistic society than other liberal democracies combined to allow more direct participation of the 1950s homophile movements in the political process and less direct confrontation between the state and homosexuality than in Germany or the English-language countries. The result has been considerable continuity in the national gay and lesbian federations of the Netherlands (as well as Denmark, Norway, and Sweden), all of which have worked well and survived from their

founding in the late 1940s and early 1950s. Gay liberation, although provoking a rethinking of the political agenda, never overturned the early organizations but instead became largely integrated into them, causing a partial name change for the COC from the Netherlands Homophile Association COC (selected in 1964) to the Netherlands Association for the Integration of Homosexuality COC in 1970–71 (see Tielman 1982; Ramsay, Heringa, and Boorsma 1974).[2] Lesbian and gay social integration has moved toward the elimination of police supervision and censorship, while the state supports access to the media, funding for social service projects and scholarly research, and legal accommodation for gay people in immigration, housing, the military, and education.

In France, the "pederastic" committee of May 1968 disappeared as quickly as it had arisen, along with the barricades of that fateful month. Not until 1971 was there a second outburst, following an issue of *Tout* (edited by Jean-Paul Sartre), that called for sexual liberation—free dis-position over one's own body, free abortion and contraception, the right to homosexuality, and the right of minors to freedom of desire (Girard 1981, 83ff; Front Homosexuel 1971). *Tout*'s "call to arms" found hundreds of adherents—as well as police seizure of the issue as an "outrage to public morals." Here emerged the Front Homosexuel Action Révolutionnaire (FHAR), which issued a *Report against Normality* (also seized by police) proclaiming a new sexual revolution. Like gay liberation, FHAR took a spontaneous turn, eschewing leadership for a series of ad hoc action groups that confronted professional "experts" and the established Left with slogans designed to explode bourgeois morality and sexual repression. The enemy was "le sexisme, le phallocratisme et l'hétérofliquisme," and FHAR declared to a startled citizenry that "we get fucked by Arabs. We're proud of it and will do it again. . . . Our asshole is revolutionary" (Girard 1981, 89–90). By 1972, FHAR had spread to major French cities, Belgium, and northern Italy as the Frente Unitario Omosessuale Rivoluzionario Italiano (FUORI), where it invaded a sexology conference in San Remo to oppose the oppressive practices of penology and psychiatry.

Again like gay liberation, FHAR soon lost its momentum, to be succeeded by a civil rights-oriented Groupe de Libération Homosexuelle (GLH), and in Italy, FUORI entered a coalition with other progressive movements in the Radical party, which took its demands to Parliament. The GLH soon split into two factions: the Groupes de Base (GLH-GB) organized in 1975 and 1976 around fighting antigay discrimination in law, employment, residence, police, and media, and the Politique et Quotidien

(GLH-PQ) which developed a more radical analysis. The first group adopted the single-issue program in an effort to bring together a broad spectrum of gay people with diverse backgrounds and beliefs. It continued to look forward to a time when social distinctions based on gender and sexual orientation could be dissolved and when the commercial ghetto would fade away unneeded. With Trotskyite inspiration, the GLH-PQ argued that homosexual identity was an invention of the bourgeoisie, the better to contain unruly desires in a police-supervised ghetto.. Why, the GLH-PQ militants wondered, were antihomosexual practices most concentrated in the institutions of repression—the family, the church, the military, the police, the prison, sports, and the schools? Might the key to a liberated society be a class struggle against the bourgeoisie (combined with the liberation of the repressed homosexuality holding together the institutions of repression? Neither of the GLH tendencies survived past 1978, but perhaps most notable was the GLH-PQ's unique development of some tenets of early gay liberation into the late 1970s.

In southern Europe and Latin America, gay organizations have proven much more ephemeral, happenstance traceable to important differences in economy and politics. Traditional gender differences, often labeled Latin machismo, have remained strong in societies where industrial employment encompasses a small portion of the population and women, especially, have not been able to enter wage labor; and thereby upset the gender system. So strong are gender codes that gender differences inscribe themselves even within homosexuality, creating two classes of men: the machos, who may with impunity take the "active" role in sex with males or females, and the effeminates (every nation has its terminology), who are stigmatized for "degrading" themselves to the status of women in bed and out (see Young 1973, 60ff.; Carrier 1976; Lacey 1979; Arboleda 1980). With a sexual semiology defined far more by gender than by sexual orientation, a gay world and identity are much less likely to develop. Lesbians, typically, have no public recognition; and the power of kin make independent same-sex relationships even less likely for women than men.

In addition, alliances between United States capitalists and indigenous landholding elites have often resulted in semifascist governments aided by successive U.S. administrations. Under such regimes, political organization of any kind becomes perilous. Notwithstanding these factors, small gay worlds have emerged in those sectors of Latin America that most resemble North America and Western Europe: in major cities with large mobile work forces that earn enough money to afford a drink in a bar. Diffusion of the gay idea clearly plays a role, as well, in the commercial establishments that consciously model themselves after American examples.

In Argentina, for example, a Frente de Liberación Homosexual formed in 1973 as part of an alignment of political forces emerging at the end of a dictatorship. Six issues of *Somos* appeared that defined a clear leftliberationist politics and included reports of the massacre of gay people under the Pinochet dictatorship in Chile. Gay liberation was forced to dissolve with the return of right-wing death squads and military government in 1976, when tens of thousands of Argentines identified with progressive movements died at their hands (see McCaskell 1976).

The first of several short-lived gay organizations began in Mexico City in 1971 when a Frente de Liberación Homosexual formed in response to the firing of several gay employees by the Sears store in Mexico City.

By the mid-1970s, gay liberation was in crisis, and out of the malaise and exhaustion after the radical phase of the gay and lesbian movement along with its New Left kin, came a reorganized and diversified set of movement groups. Most central of all the divisions that fragmented early gay liberation was that between women and men, and an autonomous lesbian feminism opened the way for revitalization.

Lesbian Feminism

> Feminism at heart is a massive complaint. Lesbianism is the solution.
> —Jill Johnston, in *Ms*.

Intense political debates and dramatic shifts in analysis characterized the emergence of lesbian activism in the early 1970s. Coming out of a flux of rapidly changing and inconsistent movement strategies developing among feminists, gay liberationists, and homophile lesbians, women went through fundamental debates about what a lesbian is and what lesbians should work for. Having a much less extensive public-bar sector than gay men have, many women came out for the first time in the midst of the women's movement and struggled for both a personal and a political orientation in an environment radically different from that of "traditional" lesbians. Because these women had so much on the line and so little anchorage in tradition, their struggles over basic questions often reached a high intensity and were resolved in frequently contradictory ways.

As late as 1970, the New York Daughters of Bilitis was holding to the cautious homophile position, only to be interrupted by the police at one meeting where they had just reaffirmed their political neutrality and had abstained from joint action with the Gay Activists Alliance. They soon reversed themselves and the DOB president, Ruth Simpson, began to invite notable feminists to speak in the ensuing months (1977; Marotta 1981). Del Martin, a DOB cofounder, had joined the National Organization for Women (NOW) in 1967, and many other lesbians were already working behind the scenes for women's rights.

Feminists at this time, however, were not always pleased to find lesbians among their ranks. Betty Friedan, in *The Feminine Mystique* (1963), had endorsed the stereotype of male homosexuality, characterizing it as "shallow unreality, immaturity, promiscuity," while leaving lesbians invisible (276). When Rita Mae Brown attempted to confront heterosexism in the women's movement in 1970 as newsletter editor of the New York chapter of NOW, Betty Friedan, then the national president of NOW, denounced a supposed "lavender menace" threatening the credibility of feminism. Brown and other suspected lesbians were purged from the organization (see Brown 1972; Abbott and Love 1972, 109–12, 127; Carden 1974, 113; Freeman 1975, 99). Similar confrontations occurred among radical feminists in Boston and at the 1971 National Women's Conference in the United Kingdom, where attempts to raise lesbian issues were rejected as "red herrings" and "private problems" (Carden 1974, 53; Walter 1980, 150).

Lesbians received a more sympathetic welcome in San Francisco in February 1970, when Gay Women's Liberation joined with the Bay Area Women's Coalition Conference. In New York, they regrouped with activists from both women's and gay liberation to hammer out the now famous manifesto, "Woman-identified Woman." Calling themselves Radicalesbians, they asserted that "a lesbian is the rage of all women condensed to the point of explosion," and pointed out that feminists could never escape the lesbian accusation. "Lesbian is the word, the label, the condition that holds women in line," they argued, "a debunking scare term that keeps women from forming any primary attachments, groups, or associations among ourselves." Lesbianism was independence from men, freedom from male approval, a matrix of women's solidarity: as such it was at the heart of feminism.

When the Second Congress to Unite Women met in New York in May, participants at a theater evening found themselves plunged into darkness. When the lights came up, they saw at the front of the auditorium twenty Radicalesbians wearing "Lavender menace" T-shirts who presented a list of grievances. The conference was liberated: workshops on lesbian issues were presented the next day, an all-women's dance was a resounding success, and the conference ended with a set of resolutions beginning, "Be it resolved that Women's Liberation is a Lesbian plot" (see Radicalesbians 1971; Teal 1971, 179–81; Hole and Levine 1971, 239–40; Abbott and Love 1972, 113–14).

But the war was not won. In the fall of 1970, when Gay People at Columbia held a public forum, Kate Millett came out as a lesbian in response to a question from the floor. *Time* magazine, which had promoted her as the preeminent feminist thinker, now announced her demise—a classic example of the tactics decried by the "Woman-identified Woman" manifesto. The case became a test of the new solidarity, and in a December press conference, leading feminists, such as Ti-Grace Atkinson, Gloria Steinem, Florynce Kennedy, Sally Kempton, Myrna Lamb, and Susan Brownmiller, rallied to Millett's defense. By 1971, even NOW had turned around, resolving that "N.O.W. acknowledges the oppression of lesbians as a legitimate concern of feminism." In 1973, at the behest of its Lesbian Caucus, the group appointed a National Task Force on Sexuality and Lesbianism (Abbott and Love 1972, 119–23, 134; Abbott 1978).

It had been an exhilarating time, which forged a major realignment of lesbian forces. "This was," remarked Jill Johnston, "a momentous series of steps from self hatred in guilt and secrecy to apologetic pleas for greater acceptance and legal sanctions to affirmation of identity to aggressive redefinition in the context of revolution" (1973, 149). The immediate outcome was a massive mobilization of lesbian energies in a cultural renaissance with the founding of such notable journals as *Ain't I a Woman?* (Iowa City), the *Furies* (Washington, D.C.), *Amazon Quarterly*, *Lesbian Tide*, *Sinister Wisdom* (Charlotte, N.C.), *Lesbian Connection* (Lansing, Mich.), *Long Time Coming* (Montreal), *Sappho* (London), and *Unsere Kleine Zeitung* (Berlin) as well as numerous local publications. A series of annual national women's music festivals began in 1973–74, stimulating an outpouring of creative talent, the rise of internationally known artists such as Meg Christian, Cris Williamson, Holly Near, and Margie Adam, and the founding of Olivia Records, devoted to the growing women's culture (St. Joan 1978; Nixon and Bergson 1978).

The redefinition of lesbianism as a form of feminist "nationalism" also spelled the end of the Daughters of Bilitis and secession from the gay movement. Both the New York chapter and the National DOB collapsed in the highly charged days of 1971. Rita Laporte and Barbara Grier seized the *Ladder* from the national DOB in 1970 to publish it as a radical lesbian journal from Reno, Nevada, but they were unable to keep it going after 1972 (see Martin and Lyon 1972, 251; Grier and Reid 1976; Marotta 1981, 263–69; D'Emilio 1983, 230).

From the beginning of gay liberation, lesbians often found themselves vastly outnumbered by men who were, not surprisingly, preoccupied with their own issues and ignorant of the concerns of women. Many women became increasingly frustrated as gay liberation men set up task groups to counter police entrapment, work for sodomy law reform, or organize dances that turned out to be 90 percent male. Men took for granted many of the social conditions that made it possible for them to be gay. But lesbians needed to address fundamental problems facing all women—such as equal opportunity in employment and violence against women—in order to have sufficient independence to become lesbian. Most men had at least the financial independence of wage labor and a well-developed commercial scene to fall back on, whereas many women were struggling to gain a foothold in employment and create places where lesbians could be together. In a movement that was supposed to forward their cause, lesbians grew angry at having to devote time and energy to "reminding" men of their existence. Many lesbians suspected that gay men would be happy to accept the place befitting their sex and class while leaving the system of male domination intact. As Marie Robertsom stated to the Canadian National Gay Rights Coalition, "Gay liberation, when we get right down to it, is the struggle for gay men to achieve approval for the only thing that separates them from the 'Man'—their sexual preference" (Robertson 1982, 177).

Early on, a move toward lesbian autonomy was underway. In April 1970, women-only dances were organized through the New York GLF to create a space where women could meet. In Los Angeles, the GLF Women's Caucus became Gay Women's Liberation and then Lesbian Feminism in rapid succession. Women-only meetings were held in CAMP-Sydney for similar reasons. With the apparent embrace of lesbianism by the women's movement in the early 1970s, lesbians around the world began withdrawing from gay liberation in 1972 and 1973. London GLF split in 1972, Lavender September formed in Amsterdam, and the Homosexuelle Aktionsgruppe Westberlin formed a Frauengruppe in 1972 (later becoming Lesbisches Aktionszentrum). Further lesbian organization in Germany usually occurred under the auspices of women's centers sponsored by feminists. Les Gouines Rouges left FHAR in Paris, and the Women's Subcommittee of the New York GAA became Lesbian Feminist Liberation.[3]

Every social movement must choose at some point what to retain and what to reject of its past. What traits and attitudes are the results of oppression and what are healthy and authentic? Which tactics come from the wisdom of forebears in facing the enemy and which merely imitate the established power system? Every movement at some time vacillates between "nationalist" and "integrationist" positions. The black movement divided over whether it wanted to affirm similarity or difference, whether it wanted to abandon Sambo-ism and claim its share of the goods of advanced capitalism, or to affirm all things African and reject a morally bankrupt and exploitative society. Gay liberation encountered similar dilemmas. Were drag queens a heterosexist stereotype acted out by self-hating homosexual men or were they the vanguard of the new gay man, rejecting the violence and misogyny of machismo and proudly coming out with their homosexuality for all to see? Feminists were not immune to the problem. Was motherhood a burden to be collectivized (or avoided) or was it women's unique contribution to humanity? Was housework merely drudgery to be shrugged off for fulfilling and *paid* employment? Was it an essential but unrecognized component in the reproduction and maintenance of the capitalist work force?

The positions taken on these questions by lesbian feminists are inextricable from the debates of the overall women's movement. After an early period of feminist integrationism, many feminists tended toward a socialist feminist camp, which argued for a comprehensive inclusion of women, gay men, and other subordinated people in a broad front against patriarchal capitalism, or toward a certain "nationalism," which aimed for a women's culture and values wherein lesbianism was revalued as the highest expression of women's solidarity and as central to women's struggle. As Ti-Grace Atkinson remarked, "Lesbianism is to feminism what the Communist Party was to the trade union movement" (1973, 14).

Many responded favorably to the new lesbian visibility, declaring themselves "political lesbians" in solidarity without necessarily involving themselves sexually with women. After the initial euphoria wore off, however, it became clear that acceptance was often superficial. Lesbian concerns were once again too often ignored, and few heterosexual women were willing to let go of their "heterosexual privilege"—what Charlotte Bunch called the "actual or promised benefits for the woman who stays in line," or "the small and short-term bribe in return for giving up lasting self-discovery and collective power" (1976, 60). This inaugurated yet another split with the development of "lesbian separatism" in 1972 to

1974. The separatists built within the nationalist position, defining lesbianism as a "woman-identified experience, . . . sharing of a rich inner life, the bonding against male tyranny, the giving and receiving of practical and political support" (Rich 1983, 192; Myron and Bunch 1975). Like other nationalisms, its theorists embarked on synthesizing a transhistorical women's mythology that reordered the universe in terms of gender opposition. Unlike the early feminists who sought to annihilate gender, insisting that such distinctions were social inventions, the nationalists adopted the opposite position, affirming an essential biological difference between men and women and working to rescue a women's culture from millennia of male domination. In contrast to male competitiveness and militarism, women would found a new civilization upon their own traditions of motherhood and nurturance.

So total was the new paradigm that Jill Johnston could claim, "considering the centrality of lesbianism to the Women's Movement it should now seem absurd to persist in associating lesbian women with the male homosexual movement. Lesbians are feminists, not homosexuals" (1975, 85). Mary Daly drew a sharp line between lesbians, whom she defined as "women who are woman-identified," and gay women, who "although they relate genitally to women, give their allegiance to men and male myths, ideologies, styles, practices, institutions, and professions." The latter group, she claimed, remained male-identified by collaborating with "heterosexist [sic] 'gay pride' protests promoted by and for men" (1978, 20, 26). If the pivotal distinction of human civilization is gender, then gay men are simply men and thus of little interest for lesbian politics.

The consolidation of lesbian identity around feminist nationalist precepts was not without problems for many lesbians. Women whose experience of lesbianism had been shaped by the bar community often found themselves rejected as "male-identified." Feminists of the early phase, who defined the core of feminism as the elimination of gender, believed that "all role playing is sick" including the "butch-fem" distinctions that remained an aspect of bar culture (Koedt 1973, 249; see Abbott and Love 1972, 36, 60; Marotta 1981, 250). Others were taken aback by the new "political lesbian" who wanted to "try it out" but knew nothing of the day-to-day hardship experienced by lesbians. Barbara Ponse found that "the self-labeled political lesbian who is bisexual or heterosexual in practice is somewhat of a mystery to women who have always defined themselves as lesbians," and many felt used by apparently bisexual women who had no interest in emotional commitment (1978, 112, 123, 212). Many suspected political lesbianism to be a form of sexual "tourism"—"the one who was going to liberate herself on my body," as Rita Mae Brown put it (1972, 191; see Gay Rev. Party 1972, 179).

Paradoxically, although the redefinition of lesbianism as a form of women's class consciousness gave permission to heterosexual women to experiment with lesbianism, it tended, at the same time, to remove sexuality from lesbian identity. As Ann Snitow, Christine Stansell, and Sharon Thompson state:

In pointing to anger rather than eros as the wellspring of lesbianism, the [Woman-identified Woman] manifesto opened the way for the desexualization of lesbian identity. . . . While the pre-feminist-movement lesbian could not forget her differences from straight women, the feminist lesbian could scarcely perceive them. Ultimately, this homogenization suppressed but could no more eliminate the tensions of difference between lesbian and straight women than it could between white women and women of color. (1983, 33)

For lesbians who decided to stay with the gay movement, feminist nationalism had taken an unfortunate turn. Whereas early feminist writers called for an end to the suppression of female sexuality, later nationalists appeared to be falling back on an image of women as above sexuality. As Jill Johnston wondered, after listening to Ti-Grace Atkinson's pleas for the political lesbian, "in her feminist rationale she had told us that the female dynamic is love and the male dynamic is sex. Translated: Man-Sex-Evil versus Woman-Love-Good" (1973, 117–18). Barbara Gittings and Kay Tobin thought that political solidarity was all very well but that the meaning of lesbianism lay elsewhere: "We believe that the majority of lesbians who come around to any gay group are not looking for analysis or warfare or reconstruction. . . . They want to meet and mix with other gay women in the legitimate pursuit of friendship and love" (1978, 151). And in an article called "Why I Am a Gay Liberationist," Chris Bearchell rejected the "imaginary world where lesbians are pure and gay men are sex perverts," arguing, "Every time a lesbian is a feminist to the world and a lesbian only to her feminist friends she is behaving with the same 'closetry' that characterized much of ghetto life, with the additional betrayal that she is doing so in the name of freedom for women" (1983, 59).

Whereas some feminists denounced the gay movement's failure to take gender abolition as its sole issue and its willingness to embrace such politically incorrect people as drag queens and butch lesbians, others viewed the movement's willingness to embrace such a diversity as a strength. And whereas Rita Mae Brown and Martha Shelley were attacked as male-identified for promoting coalition between lesbians and people oppressed by class and race, others believed that social transformation could not be a question of gender alone. Gittings and Tobin claimed that lesbian separatists had identified the wrong enemy with their "supercharged response to sexism and male chauvinism, to the point that they spend much time and energy attacking the sexism of the handiest men around, the gay men in the movement" (1978, 151). And Bearchell complained:

It is, after all, our sexuality, and the sexual minorities in our community, that are under attack. Here, it seems, is where we must defend ourselves. But suddenly the same radical feminists who had denounced gay liberationists for our concern with such un-radical things as rights, were nervous about being in a coalition with us because we might take some not-quite-respectable position on sex. (1983, 59)

Gay men's reactions to the lesbian secession ranged from breast-beating to confusion and resentment. About the time many feminists became political lesbians, some gay men became "effeminists," taking to heart lesbian criticism of their male privilege and renouncing all personal signs of masculinity. The effeminists recalled the earlier debates over drag in gay liberation and later reemerged as "radical faeries" searching for a tradition of "gay male spirituality" parallel to feminist cultural nationalism. Henry Hay, a founder of the first Mattachine and his lover, John Burnside, figured among its relatively small number of adherents (Collier and Ward 1980; Hardy 1980). Most gay organizations scrambled—often too late—to accommodate lesbian demands, but some groups successfully retained female and male participation by moving toward parity decision making in the organization's day-to-day affairs. Other gay men suspected that many lesbians simply found it more convenient to identify with a large and respectable mass movement than with a group of stigmatized "perverts." After all, feminism offered quick rehabilitation to a reviled sexual identity, and many men were appalled at the willingness of newly respectable lesbians to denounce the lives of gay men as no more than "economic and cultural privilege, . . . anonymous sex . . . pederasty . . . ageism" (Rich 1983, 192).

With the feminist movement continuing to develop inconsistent trends in the late 1970s, debates on the future of lesbian organization went on unabated. Cultural nationalism offered quite a different agenda for lesbian struggle than did integrationist approaches, whether from a liberal civil rights approach or from a more radical socialist feminist model. The immediate outcome of these conflicts has been considerable fragmentation of lesbian energies and a renewed confrontation between these tendencies in the sex debates of the 1980s. But for this story, we must look at the movement in the 1980s.

The Movement and the Grass Roots

The paradox of the 1970s was that gay and lesbian liberation did not produce the gender-free communitarian world it envisioned, but faced an unprecedented growth of gay capitalism and a new masculinity. While debates raged inside the movement, the actions of gay liberationists and lesbian feminists entered a larger political field, which transformed and expanded the gay world in unexpected directions.

The most immediate effect of the movement upon the masses of gay men and lesbians, who were largely unacquainted with its internal debates and struggles, was a new sense of pride, an honest affirmation of a personal emotional life, a sense of relief at not having always to hide or apologize, and a new claim (or reclamation) of the symbols of masculinity. After the "gender-fuck" drag of the early 1970s and the intense critique of gender, both lesbians and gay men began more and more to embody a certain working-class ideal of masculinity; the fashion was "jeans and denim workmen's overalls . . . topped by a man's T-shirt or workshirt . . . [and] heavy men's workboots or sneakers" (Cassell 1972, 83). While heterosexual men were relaxing into a new androgyny in the 1970s, adopting longer hair, brighter colors, and softer fabrics, gay men and lesbians were making a mass commitment to denim, plaid, and leather.

The gender shift is perhaps not so surprising in retrospect. Masculine symbols offer the most ready-at-hand vocabulary of self-assertion. As gay people gained self-confidence and demanded respect, they began to present themselves as serious and tough. Still, it is important not to confuse this artful masculinity with conventional male chauvinism. The new gay masculinity had a specific meaning. It was an open secret among gay men that the apparent motorcyclists and cowboys standing in gay bars were gentle men at heart. Although this cultivated masculinity was a disappointment to gay men of the old school who thought they wanted "real men," for most the apparent inconsistency was attractive and right. Among lesbians, the new dress code rejected the incapacitating delicacy and frilliness vaunted by the heterosexist press in favor of the self-reliant image of the Amazon. Once the new self-confidence was fully internalized, the masculinist style began to wane—but more quickly for lesbians than for gay men.

In another sense, the new masculinity participated in one of the deepest aspirations of the movement, that is, to develop egalitarian relationships free from role playing. In this, the movement was an inheritor of a two-hundred-year trend toward egalitarian ideals in the companionate marriage. Long the victims of male violence and control in families, women had sought to improve their status at home and espouse full equality. It might be argued that homosexual relationships have an inherent interest in shedding gender and that they have, in fact, pioneered work sharing and role flexibility in coupled relationships. Whether ahead of or with progressive trends in heterosexual relationships, Stonewall marked a decisive break with a waning tradition of gender within homosexuality. The masculinization of the 1970s dissolved remnants of the "real man" versus "queer" distinction (described above as the Latin American model), which is so evident in historical documents of gay life in the West. As Rudy Kikel put it: "Up until liberation, I really feel that we were all in love with straight men. . . . what we found was that we could find that [maleness] in each other. And the great benefit was that we became sexual objects for each other" (1981, 12). As well, sex roles (as opposed to gender) largely disappeared in the 1970s in that "the most common set of sexual preferences among gay males is for all roles, both oral and anal and active and passive" (Harry 1976–77, 150). Joseph Harry found only a folk distinction between "versatile" and "not versatile" but not a distinction between sex roles.[4]

None of this is to say that the new trends solved the much thornier micropolitics of day-to-day living or that gender entirely lost its meanings. As women have increasingly entered male-identified jobs and vice versa (and gay people are on the cutting edge of this change), gender has become more and more disarticulated from the division of labor. Still, problems of initiative and response, active and passive, domination and submission, continue to crop up in actual relationships, and discussions of them inevitably become entangled in gender vocabularies, which have so long characterized the differences. In the gay world, drag has been shunted off to the side, becoming a "little tradition" outside a larger main-

stream. Transvestism has become a world of its own with many female impersonators developing professional identities and straight audiences. Debates among lesbians about butch-fem relationships have resurfaced to recover what was valuable in the bar dyke tradition and to rearticulate feelings and practices that egalitarian slogans never dealt with.

The other paradoxical outcome of gay liberation was the expansion of the gay ghetto. The success of the movement in beating back state management and repression of gay places allowed for a new generation of businesses oriented to a gay market. Within a decade, every major city in North America and Western Europe had a new range of bars and saunas, restaurants and discos, travel agents and boutiques, lawyers and life insurers, social services and physicians, who catered specifically to a gay clientele. At the same time, there was a remarkable development of many women-owned and operated places, many of which were havens for lesbians.

While gay liberation zapped public institutions, a new class of small businessmen (and some women) began carving out a commercial ghetto that directly touched the lives of many more gay people than the movement itself. While gay liberation theory presumed that the release of homosexuality would explode conventional sexual and familial arrangements, capitalist environments cultivated new institutions compatible with itself. The result, remarked Dennis Altman, was a new masculine gay man who was "non-apologetic about his sexuality, self-assertive, highly consumerist and not at all revolutionary, though prepared to demonstrate for gay rights" (Altman 1980, 52).

The capitalization of homosexuality in the 1970s shaped gay male identity in quite another way. Neither Ginsberg's vision of "tender lovers and fellows" nor gay liberation's democratic gay community could come to pass under such auspices. As businessmen developed efficient sex delivery systems for gay men, a world of adhesive comrades and brothers became a more remote ideal. The unique potential inherent in homosexuality to rehumanize relationships among men became increasingly closed off in favor of orgasm without communication. Relationships among men were participating in the growing sexual reductionism of the current century: male bonding in the commercial gay world tended to implode into its sexual aspect, and those who did manage to make long-term commitments to each other often withdrew from the commercial world to do so. As Laud Humphreys observed: "In the *Wealth of Nations*, Adam Smith postulated the ideal form of human relationship as being specific, deper-

The Rise of a Gay and Lesbian Movement

sonalized, short-term and contractual. This capitalist ideal is realized in the sex exchange of the homosexual underworld [sic]" (1972a, 66)

The new sexual "freedom" brought a tremendous release of energy and profound exploration of erotics as a value in itself. Early gay and lesbian liberationists looked forward to an era of plural bonding, freed of the oppressive weight of monogamy, jealousy, and sexual boredom. But many gay men began to feel a certain sexual alienation and emotional suffocation. As Robert Patrick bitterly recalled:

Bars wouldn't make any money if people loved one another. You go to a bar, get drunk, get drugged up and go into the back room and have your sex there. The economic reality is that your cock is being sold by that bar. Your ass is being sold by that bar. But we don't ever think of it that way. (1979, 8)[5]

Thus, whereas the lesbian movement began to submerge sexual topics under talk of sisterhood, thereby slipping back toward traditional definitions of female sexuality, gay men found themselves unable to talk forthrightly of their need for love, confirming traditional male socialization that demands that men be sexual but unemotional. As Andrew Holleran confessed in a perceptive article for *Christopher Street*:

Last week in the baths I was sitting in a corner waiting for Mister Right when I saw two men go into an even darker nook and run through the entire gamut of sexual acts. And when they were finished—after all these *kisses* . . . and *moans* and *gasps*, things that caused scandals in the nineteenth century, toppled families, drove Anna Karenina to suicide—. . . after all that, they each went to a separate bedroom to wash up. Now you may view this as the glory of the zipless fuck, but I found it suddenly—and it surprised me, for I'd always adored this event before—the most reductive, barren version of sex a man could devise. (1979, 12)

The commercial gay world could provide "fast-food" sex, but it did nothing to nurture lasting relationships among men. It contained and marketed gay male sexuality back to gay men, but reproduced the competitive alienation among men experienced in the larger society. It was, in fact, as Joseph Harry and William Devall found, a satisfactory arrangement for "persons with significant components of heterosexuality in their self-identity" who "vacationed" in the gay ghetto but had no interest in emotional involvement with other men (1978).[6] But for gay men, it was not always enough.

The irony of the 1970s, then, was the ease with which gay and lesbian aspirations were assimilated, contained, and overcome by the societies

Gay Liberation and Lesbian Feminism

in which they originated. The gender challenge of the liberation movements (itself imminent in the increasingly complex division of labor of modern capitalism) became the gender affirmation of the end of the decade, whether as gay male masculinity or lesbian feminist nationalism. The socialist challenge of the New Left helped contribute to its opposite: a bigger commercial ghetto. Still, these paradoxes were not simply historical cycle: or pendulum swings against an unchanging background. Each social onvulsion pulled out, amplified, and rewove disparate discursive strads into different social patterns. Each stage experimented with new combinations of received elements, producing a changed social fabric. But tie disarray of outcomes was soon to fall prey to a reorganized enemy as conservative forces in the United States formed the New Right.

19

Homophobia and Women's Sport

ELAINE M. BLINDE AND DIANE E. TAUB

Blinde and Taub explore the role of attributed sexual orientation in disempowering women who violate gender norms: varsity female collegiate athletes. By challenging the gender order and opposing male domination, these women intrude into a traditional male sanctum and threaten the male domain of physicality and strength. By casting the lesbian label on women athletes, society stigmatizes them as masculine and as sexual perverts. While the homosexual label is routinely used to degrade male athletes who fail to live up to the hyper-masculine ideal, the lesbian label is used to divide and silence female athletes. They may adopt the perspective of their oppressors and demean their teammates as lesbians, thus destroying team solidarity, and/or shun the label, but be forced to acknowledge its demeaning power as they attempt to escape it. The forceful effect of the lesbian label applied to women athletes shows the dominance not only of heterosexuals over homosexuals, but of men over women.

Central to the preservation of a patriarchal and heterosexist society is a well-established gender order with clearly defined norms and sanctions governing the behavior of men and women. This normative gender system is relayed to and installed in members of society through a pervasive socialization network that is evident in both everyday social interaction and social institutions (Schur 1984). Conformity to established gender norms contributes to the reproduction of male dominance and heterosexual privilege (Lenskyj 1991; Stockard and Johnson 1980).

Despite gender role socialization, not all individuals engage in behavior consistent with gender expectations. Recognizing the potential threat of such aberrations, various mechanisms exist that encourage compliance with the normative gender order. Significant in such processes are the stigmatization and devaluation of those whose behavior deviates from the norm (Schur 1984).

Women's violation of traditional gender role norms represents a particularly serious threat to the patriarchal and heterosexist society because this deviant behavior resists women's subordinate status (Schur 1984). When women

---

"Homophobia and Women's Sport: The Disempowerment of Athletes," by Elaine M. Blinde and Diane E. Taub, *Sociological Focus*, Vol. 25, No. 2, May 1992. Reprinted by permission.

engage in behavior that challenges the established gender order, and thus opposes male domination, attempts are often made by those most threatened to devalue these women and ultimately control their actions. One means of discrediting women who violate gender norms and thereby questioning their "womanhood" is to label them lesbian (Griffin 1987).

The accusation of lesbianism is a powerful controlling mechanism given the homophobia that exists within American society. Homophobia, representing a fear of or negative reaction to homosexuality (Pharr 1988), results in stigmatization directed at those assumed to violate sexuality norms. Lesbianism, in particular, is viewed as threatening to the established patriarchal order and heterosexual family structure since lesbians reject their "natural" gender role, as well as resist economic, emotional, and sexual dependence on men (Gartrell 1984; Lenskyj 1991).

As a means for both discouraging homosexuality and maintaining a patriarchal and heterosexist gender order (Pharr 1988), homophobia controls behavior through contempt for purported norm violators (Koedt, Levine, and Rapone 1973). One method of control is the frequent application of the lesbian label to women who move into traditional male-dominated fields such as politics, business, or the military (Lenskyj 1991). This "lesbian baiting" (Pharr 1988:19) suggests that women's advancement into these arenas is inappropriate. Such messages are particularly potent since they are lodged in a society that condemns, devalues, oppresses, and victimizes individuals labeled as homosexuals (Lenskyj 1990).

Another male arena in which women have made significant strides, and thus risk damaging accusation and innuendo, is that of sport (Blinde and Taub 1992; Lenskyj 1990). Sport is a particularly susceptible arena for lesbian labeling due to the historical linkage of masculinity with athleticism (Birrell 1988). When women enter the domain of sport they are viewed as violating the docile female gender role and therefore extending culturally constructed boundaries of femininity (Cobhan 1982; Lenskyj 1986; Watson 1987). The attribution of masculine qualities to women who participate in sport leads to a questioning of their sexuality and subsequently makes athletes targets of homophobic accusations (Lenskyj 1986)....

Therefore, the present study explores the stereotyping of women athletes as lesbians and the accompanying homophobia fostering this label. General themes and processes which inform us of how these individuals handle the lesbian issue are identified. These dynamics are grounded in the contextual experiences of women athletes and relayed through their voices.

Athletic directors at seven large Division I universities were contacted by telephone and asked to participate in a study examining various aspects of the sport experience of female college athletes. These administrators were requested to provide a list of the names and addresses of all varsity women athletes for the purpose of contacting them for telephone interviews.... Interested athletes were encouraged to return an informed consent form indicating their willingness to participate in a tape-recorded telephone interview. Based on this initial contact, a total of 16 athletes agreed to be in the study.

In order to increase the sample size to the desired 20 to 30 respondents, lected from the three lists. Eight of these athletes agreed to be interviewed, resulting in a final sample size of 24. Athletes in the sample were currently participating in a variety of women's inter-collegiate varsity sports—basketball (n = 5), track and field (n = 4), volleyball (n = 3), swimming (n = 3), softball (n = 3), tennis (n = 2), diving (n = 2), and gymnastics (n = 2). With an average age of 20.2 years and overwhelmingly Caucasian (92%), the sample contained 2 freshmen, 9 sophomores, 5 juniors, and 8 seniors. A majority of the athletes (n = 22) were recipients of an athletic scholarship....

Semi-structured telephone interviews were conducted by two trained female interviewers. All interviews were tape-recorded and lasted from 50 to 90 minutes. Questions were open-ended in nature so that athletes would not feel constrained in discussing those issues most relevant to their experiences. Follow-up questions were utilized to probe how societal perceptions of women athletes impact their behavior and experiences.

RESULTS

Examination of the responses of athletes revealed two prevailing themes related to the presence of the lesbian stereotype in women's sport—(a) a silence surrounding the issue of lesbianism in women's sport, and (b) athletes' internalization of societal stereotypes concerning lesbians and women athletes. It is suggested that these two processes disempower women athletes and thus are counterproductive to the self-actualizing capability of sport participation (Theberge 1987).

SILENCE SURROUNDING LESBIANISM IN WOMEN'S SPORT

One of the most pervasive themes throughout the interviews related to the general silence associated with the lesbian stereotype in women's sport. Although a topic of which athletes are cognizant, reluctance to discuss and address lesbianism in women's sport was evident. Based on the responses of athletes, this silence was manifested in several ways: (a) athletes' difficulty in discussing lesbian topic, (b) viewing lesbianism as a personal and irrelevant issue, (c) disguising athletic identity to avoid lesbian label, (d) team difficulty in addressing lesbian issue, and (e) administrative difficulty in addressing lesbian issue.

ATHLETES' DIFFICULTY IN DISCUSSING LESBIAN TOPIC

Initial indication of silencing was illustrated by the difficulty and uneasiness many athletes experienced in discussing the lesbian stereotype. Some respondents were initially reluctant to mention the topic of lesbianism; discussion feelings of uneasiness or discomfort. Athletes were most likely to introduce this topic when questions were asked about societal perceptions of women's sport and female athletes, as well as inquiries about the existence of stereotypes associated with women athletes. Moreover, the lesbian issue was sometimes discussed without specifically using the term lesbian. For example, some athletes evaded the issue by making indirect references to lesbianism (e.g., using the word "it" rather than a more descriptive term)....

Respondents' approach to the topic of lesbianism indicates the degree to which women athletes have been socialized into a cycle of silence. Such silence highlights the suppressing effects of homophobia. Moreover, athletes' reluctance to discuss topics openly related to lesbianism may be to avoid what Goffman (1963) has termed "courtesy stigma," a stigma conferred despite the absence of usual qualifying behavior.

VIEWING LESBIANISM AS A PERSONAL AND IRRELEVANT ISSUE

A second indicator of the silence surrounding the lesbian stereotype was reflected in athletes' general comments about lesbianism. Many respondents indicated that sexual orientation was a very personal issue and thus represented a private and extraneous aspect of an individual's life. These athletes felt it was inappropriate for others to be concerned about the sexual orientation of women athletes.

Although such a manifestation of silence might reflect the path of least resistance by relieving athletes of the need to discuss or disclose their sexual orientation (Lenskyj 1991), it does not eliminate the stigma and stress experienced by women athletes. Also, making lesbianism a private issue does not confront or challenge the underlying homophobia that allows the label to carry such significance. The strategy of making sexual orientation a personal issue depoliticizes lesbianism and ignores broader societal issues.

DISGUISING ATHLETIC IDENTITY TO AVOID LESBIAN LABEL

A third form of silence surrounding the lesbian stereotype was the tendency for athletes to hide their athletic identities. Nearly all respondents indicated that despite feeling pride in being an athlete, there were situations where they preferred that others not know their athletic identity. Although not all athletes indicated that this concealment was to prevent being labeled a lesbian, it was obvious that there was a perceived stigma associated with athletics that many women wanted to avoid (e.g., masculine women, women trying to be men, jock image). In most cases, respondents indicated that disguising their athletic identity was either directly or indirectly related to the lesbian stereotype....

Athletes also stated that they (or other athletes they knew) accentuated certain behaviors in order to reduce the possibility of being labeled a lesbian. Being seen with men, having a boyfriend, or even being sexually promiscuous with men were commonly identified strategies to reaffirm an athlete's heterosexuality. As one athlete commented, "If you are a female athlete and do not have a boyfriend, you are labeled [lesbian]."

As reflected in the responses of athletes, the role of sport participant was often intentially de-emphasized in order to reduce the risk of being labeled lesbian. Modification of athletes' behavior, even to the point of denying critical aspects of self, was deemed necessary for protection from the negativism attached to the lesbian label. This disguising of athletic identity exemplifies what Kitzinger (1987:92) termed "role inversion." In such a situation individuals attempt to demonstrate that their group stereotype is inaccurate by accentuating traits that are in opposition to those commonly associated with the group (in the case of women athletes, stressing femininity and heterosexuality).

TEAM DIFFICULTY IN ADDRESSING LESBIAN ISSUE

Not only did the silence surrounding lesbianism impact certain aspects of the lives of individual athletes, but it also affected interpersonal relationships among team members. This silence was often counterproductive to the development of positive group dynamics (e.g., team cohesion, open lines of communication).

As was often true at the individual level, women's sport teams were unable collectively to discuss, confront, or challenge the labeling of women athletes as lesbians. One factor complicating the ability of women athletes to confront the lesbian stereotype was the divisive nature of the label itself (Gentile 1982); the lesbian issue sometimes split teams into factions or served as the basis for clique formation.

Heterosexual and lesbian athletes often had limited interaction with each other outside the sport arena. Moreover, athletes established distance between themselves and those athletes most likely to be labeled lesbian (i.e., those possessing "masculine" physical or personality characteristics)....

From the interviews, there was little evidence that lesbian and nonlesbian athletes collectively pooled their efforts to confront or challenge the lesbian stereotype so prevalent in women's sport. The silence surrounding lesbianism creates divisions among women athletes; this dissension has the effect of preventing female bonding and camaraderie (Lenskyj 1986). Rather than recognizing their shared interests, women athletes focus on their differences and thus deny the formation of "alliance" (Pheterson 1986:149). This difficulty in attaining team cohesion is unfortunate since women's sport is an activity where women as a group can strive for common goals (Lenskyj 1990). The lesbian stereotype not only limits female solidarity, but also minimizes women's ability to challenge collectively the patriarchal and heterosexist system in which they reside (Bennett et al. 1987).

1986). Finally, as a result of this preoccupation with silence, women athletes often engage in self-denial as they hide their athletic identity.

ATHLETES' INTERNALIZATION OF SOCIETAL STEREOTYPES

A second major theme reflected in the responses of athletes was a general internalization of stereotypic representations of lesbians and women athletes. As argued by Kitzinger (1987) and Pheterson (1986), members of oppressed and socially marginalized groups often find themselves accepting the stereotypes and prejudices held by the dominant society. Representing "internalized oppression" (Pheterson 1986:148), the responses of athletes revealed an identification with the aggressor, self-concealment, and dependence on others for self-definition (Kitzinger 1987; Pheterson 1986). Acceptance of these societal representations by a disadvantaged group (in this case women athletes) grants legitimacy to the position of those who oppress and contributes to the continued subordination of the oppressed (Wolf 1986). Based on our interviews, athletes' internalization of stereotypes and prejudices were reflected by three categories of responses: (1) acceptance of lesbian stereotypes, (2) acceptance of women's sport team stereotypes, and (3) acceptance of negative images of lesbianism.

ACCEPTANCE OF LESBIAN STEREOTYPES

In response to various open-ended questions, it was apparent that athletes were able to identify a variety of factors that they felt led others to label women athletes as lesbians (e.g., physical appearance, dress, personality characteristics, nature of sport activity). Given that the attribution of homosexuality is most likely to be associated with traits and behaviors judged to be more appropriate for members of the opposite sex (Dunbar, Brown and Amoroso 1973; Dunkle and Francis 1990), it was not surprising that athletes' rationale for the lesbian label included such attributes as muscularity, short hair, masculine clothing, etc.

When athletes were asked about the validity of the lesbian label in women's sport, affirmative replies were frequently based on conjecture. For example, to provide support for why they felt there was a basis for labeling women athletes as lesbians, respondents made such comments as "there are masculine girls on some teams," "it is really obvious," or "you can just tell that some athletes are lesbians."

These explanations tend to reflect an acceptance of societal definitions of lesbianism—beliefs that are largely male-centered and supportive of a patriar-

ADMINISTRATIVE DIFFICULTY IN ADDRESSING LESBIAN ISSUE

Another manifestation of silence relayed in the responses of athletes was the apparent unwillingness of coaches and athletic directors to confront openly the lesbian stereotype. As was found with individual athletes and teams, those in leadership positions in women's sport refused to address or challenge this stereotype. Reluctance to confront the lesbian issue at the administrative level undoubtedly influenced the manner in which athletes handled the stereotype....

Because the women's intercollegiate sport system is homophobic and predominately male-controlled (i.e., over half of coaches and four-fifths of administrators are men) (Acosta and Carpenter 1992), it is assumed that survival in women's sport requires collusion in a collective strategy of silence about and denial of lesbianism (Griffin 1987). Coaches and administrators fear that openly addressing the lesbian issue may result in women's sport losing the recent gains made in such areas as fan support, budgets, sponsorship, and credibility (Griffin 1987). Therefore, leaders yield to this fear as they strive to achieve acceptability for women's sport. Such accommodation to the patriarchal, heterosexist sport structure not only contributes to isolation as coaches and administrators are afraid to discuss lesbianism, but also limits their identification with feminist and women's issues (Duquin 1981; Hargreaves 1990; Pharr 1988; Zipter 1988).

SUMMARY

Based on athletes' responses, it was evident that the silence surrounding the lesbian issue in women's sport was deeply ingrained at all levels of the women's intercollegiate sport structure. Such widespread silencing reflects the negativism and fear associated with lesbianism that are so prevalent in a homophobic society. This strategy of silence or avoidance, however, is counterproductive to efforts to dispel or minimize the impact of the lesbian stereotype. Not only does silence disallow a direct confrontation with those who label athletes lesbian, but it also perpetuates the power of the label by leaving unchallenged rumors and insinuations. Moreover, the fear, ignorance, and negative images that are frequently associated with women athletes are reinforced by this silence (Zipter 1988).

Numerous aspects of women's experience in sport are ignored due to the silence surrounding the subject of lesbianism. For example, refusing to address this issue has limited understanding of the dimensionality and complexity of women's sport participation. Moreover, since the stigma associated with the lesbian label inhibits athletes from discussing this topic with each other, these women frequently do not realize that they possess shared experiences that would provide the foundation for female bonding. Without an "alliance" research has shown that people associate physical appearance with homosexuality (Levitt and Klassen 1974; McArthur 1982; Unger, Hilderbrand, and Madar 1982). For example, attractiveness is equated with heterosexuality and a larger, muscular body build is identified with lesbianism.

Moreover, the remarks of athletes demonstrate that the very group that is oppressed (in this case women athletes) accepts societal stereotypes about lesbians and has incorporated these images into their managing of the situation. As suggested by Garrell (1984) and certainly evident in this sample of women athletes, cultural myths about lesbianism perpetuated in a homophobic society are often firmly ingrained in the thinking of affected individuals.

ACCEPTANCE OF WOMEN'S SPORT TEAM STEREOTYPES

Relative to providing a rationale for why the lesbian label was more likely to be associated with athletes in certain sports, respondents again demonstrated an understanding and internalization of societal stereotypes. The sports most commonly identified with the lesbian label were softball, field hockey, and basketball. In attempting to explain why these team sports were singled out, athletes mentioned such factors as the nature of bodily contact or amount of aggression in the sport, as well as the body build, muscularity, or athleticism needed to play the sport.

Respondents often relied on the "masculine" and "feminine" stereotypes to differentiate sports in which participating women were more or less likely to be subjected to the lesbian label. Although participants in team sports were more likely than individual sports (e.g., gymnastics, swimming, tennis, golf) to be associated with the lesbian label, it was interesting to note that volleyball was often exempt from the connotations of lesbianism.

The higher incidence of lesbian labeling found in team sports (as opposed to individual sports) may also be related to the potential that team sports provide for interpersonal interactions. As previously mentioned, emphasizing teamwork and togetherness, team sports allow women rare opportunities to bond collectively in pursuit of a group goal (Lenskyj 1990). Recognition of this power of female bonding is often reflected by male opposition to women-only activities (Lenskyj 1990).

ACCEPTANCE OF NEGATIVE IMAGES OF LESBIANISM

During the course of the interviews, a large majority of athletes made comments about lesbians which reflected an internalization of the negativism associated with lesbianism. Respondents also demonstrated a similar acceptance when they relayed conversations they had had with both teammates and outsiders.

One form of negativism was reflected by statements that specifically "put down" lesbians. Athletes' negative comments about lesbians were included in conversations with outsiders so others would not associate the lesbian label with them. Representing a form of projection (Gross 1978), some athletes attempted to dissociate from traits that they saw in themselves (e.g., strength, muscularity, aggressiveness)....

It is ironic that athletes rarely directed their anger or condemnations at the homophobic society that restricts the actions of women athletes, including the nonlesbian athlete. Rather, by focusing on athletes as lesbians, a blame the victim approach diverts attention from the cause of the oppression (Pharr 1988). As is often true of oppressed groups, a blame the victim philosophy results in an acceptance of the belief system of the oppressor (in this case a patriarchal, heterosexist society) (Pharr 1988). Like other marginalized groups, women athletes accept the normative definitions of their deviance (Kitzinger 1987); in effect, such responses represent a form of collusion with the oppressive forces (Pheterson 1986). Interestingly, no mention was made by respondents about attempts to engage the assistance or support of units on campus sympathetic to gay and lesbians issues (e.g., feminist groups, gay and lesbian organization, affirmative action offices).

SUMMARY

From the interview responses, it was evidence that athletes had internalized societal stereotypes related to lesbians and women athletes, as well as the negativism directed toward lesbianism. This acceptance was so ingrained in these athletes that they were generally unaware of the political ramifications of both lesbianism and the accompanying lesbian stereotype as applied to women athletes. Despite their gender norm violation as athletes, these women often had a superficial understanding of gender issues. Such a lack of awareness may be due in part to the absence of a feminist consciousness in athletes (Boutilier and SanGiovanni 1983; Kaplan 1979) and their open disavowal of being a "feminist," "activist," or "preacher of women's liberation." Accepting societal definitions of their deviance, as well as the inability to see their personal experiences as political in nature, attests to this limited consciousness (Boutilier and SanGiovanni 1983). Athletes' responses are indicative of the degree to which they exhibit internal homophobia so common in American society.

Only a few athletes possessed deeper insight into factors that may underlie the labeling of women athletes as lesbians. For example, one respondent felt women athletes were a "threat to men since they can stand on their own feet." Or, in another situation, an athlete viewed lesbian labeling as a means to devalue women athletes or successful women in general. Still another respondent suggested the label stemmed from jealousy and thus was used as a means to "get back" at women athletes. These rare remarks by respondents transcend the blame the victim view held by the majority of athletes. Such commitments indicate a deeper understanding of how homophobia and patriarchal ideology limit or control women's activities and their bodies.

DISEMPOWERMENT

Given the silence surrounding the lesbian issue and the degree to which athletes have internalized societal images of lesbians and women athletes, the presence of the lesbian stereotype has negative ramifications for women athletes. Although sport participation possesses the potential for creativity and physical excellence (Theberge 1987), women modify their behavior so they will not be viewed as "stepping out of line." Women athletes become disempowered (Pharr 1988) through processes that detract from or reduce the self-actualizing potential of the sport experience.

Attaching the label of lesbian to women who engage in sport diminishes the sporting accomplishments of athletes. Women athletes are seen as something less than "real women" because they do not exemplify traditional female qualities (e.g., dependency, weakness, passivity); thus their accomplishments are not viewed as threatening to men (Birrell 1988). Interestingly, the athlete interviewed believed that the specific group most likely to engage in lesbian labeling was male athletes.

Discrediting women with the label of lesbian works further to control the number of females in sport, particularly in a homophobic society where prejudice against lesbians is intense (Birrell 1988; Zipter 1988). Keeping women out of sport, in turn, prevents females from discovering the power and joy of their own physicality (Birrell 1988) and experiencing the potential of their body. Moreover, discouraging women from participating in sport disempowers them by removing an arena where women can bond together (Birrell 1988; Cobhan 1982)....

Another form of disempowerment occurs for those athletes who are lesbians. Intense homophobia often forces lesbians to deny their very essence, thus making the lesbian athlete invisible. Concealment, although protecting the lesbian athletes' identity, imposes psychological strain and can undermine positive self-conceptions (Schur 1984). Misrepresenting their sexuality, lesbian athletes are not in a position to confront the homophobia so prevalent in women's sport. Consequently, this ideology not only remains intact, but also is strengthened (Ettore 1980).

REFERENCES

Acosta, R. Vivian and Linda Jean Carpenter. 1992. "Women in Intercollegiate Sport: A Longitudinal Study—Fifteen Year Update 1977–1992." Unpublished manuscript, Brooklyn College, Department of Physical Education, Brooklyn.

Bennett, Roberts S., K. Gail Whitaker, Nina Jo Woolley Smith, and Anne Sablove. 1987. "Changing The Rules of The Game: Reflections Toward A Feminist Analysis of Sport." *Women's Studies International Forum* 10: 369–386.

Birrell, Susan. 1988. "Discourses on The Gender/Sport Relationship: From Women in Sport to Gender Relations." Pp. 459–502 in *Exercise And Sport Science Reviews*, vol. 16, edited by K. B. Pandolf. New York: MacMillan.

Blinda, Elaine M. and Diane E. Taub. 1992. "Women Athletes as Falsely Accused Deviants: Managing the Lesbian Stigma." *The Sociological Quarterly*.

Boutilier, Mary A. and Lucinda SanGiovanni. 1983. *The Sporting Woman*. Champaign, IL: Human Kinetics.

Cobban, Linn Ni. 1982. "Lesbians in Physical Education And Sport. Pp. 179–186 in *Lesbian Studies: Present And Future*, edited by M. Cruikshank. New York: Feminist Press.

Dunbar, John, Marvin Brown and Donald M. Amoroso. 1973. "Some Correlates of Attitudes Toward Homosexuality." *Journal of Social Psychology* 89: 271–279.

Dunkle, John H. and Patricia L. Francis. 1990. "The Role of Facial Masculinity/Femininty in The Attribution of Homosexuality." *Sex Roles* 23: 157–167.

Duquin, Mary E. 1981. "Feminism And Patriarchy in Physical Education." Paper presented at the annual meetings of the North American Society for the Sociology of Sport, Fort Worth, TX.

Ettore, E. M. 1980. *Lesbians, Women and Society*. London: Routledge

Gartrell, Nanette. 1984. "Combating Homophobia in The Psychotherapy of Lesbians." *Women And Therapy* 3: 13–29.

Gentile, S. 1982. "Out of The Kitchen." *City Sports Monthly* 8: 27.

Goffman, Erving. 1963. *Stigma: Notes on The Management of Spoiled Identity*. Englewood Cliffs, NJ.: Prentice-Hall.

Griffin, Patricia S. 1987. "Homophobia, Lesbians, And Women's Sports: An Exploratory Analysis." Paper presented at the annual meetings of the American Psychological Association, New York.

Gross, Martin L. 1978. *The Psychological Society*. New York: Simon and Schuster.

Hargreaves, Jennifer A. 1990. "Gender on The Sports Agenda." *International Review for Sociology of Sport* 25: 287–308.

Kaplan, Janice. 1979. *Women And Sports*. New York: Viking.

Kitzinger, Celia. 1987. *The Social Construction of Lesbianism*. London: Sage.

Koedt, Anne, Ellen Levine and Anita Rapone. 1973. *Radical Feminism*. New York: Quadrangle.

Lenskyj, Helen. 1986. *Out of Bounds: Women, Sport And Sexuality*. Toronto: Women's Press.

———. 1990. "Power And Play: Gender And Sexuality Issues in Sport and Physical Activity." *International Review for Sociology of Sport* 25: 235–245.

———. 1991. "Combating Homophobia in Sport And Physical Education." *Sociology of Sport Journal* 8: 61–69.

Levit, Eugene E. and Albert D. Klassen, Jr. 1974. "Public Attitudes toward Homosexuality: Part of the 1970 National Survey by The Institute for Sex Research." *Journal of Homosexuality* 1: 29–43.

McArthur, Leslie Z. 1982. "Judging A Book by Its Cover: A Cognitive Analysis of The Relationship between Physical Appearance And Stereotyp-

Psychology, edited by Albert H. Hastorf and Alice M. Isen. New York: Elsevier/North-Holland.

Pharr, Suzanne. 1988. *Homophobia: A Weapon of Sexism.* Inverness, CA: Clurdon.

Pheterson, Gail. 1986. "Alliances between Women: Overcoming Internalized Oppression And Internalized Domination." *Signs: Journal of Women in Culture And Society* 12: 146–160.

Schur, Edwin M. 1984. *Labeling Women Deviant: Gender, Stigma, And Social Control.* New York: McGraw-Hill.

Stockard, Jean and Miriam M. Johnson, 1980. *Sex Roles: Sex Inequality And Sex Role Development.* Englewood Cliff, NJ: Prentice-Hall.

Theberge, Nancy. 1987. "Sport And Women's Empowerment." *Women's Studies International Forum* 10: 387–393.

Unger, Rhoda K., Marcia Hilderbrand and Theresa Madar. 1982. "Physical Attractiveness And Assumptions about Social Deviance: Some Sex-By-Sex Comparisons." *Personality And Social Psychology Bulletin* 8: 293–301.

Watson, Tracey. 1987. "Women Athletes And Athletic Women: The Dilemmas And Contradictions of Managing Incongruent Identities." *Sociological Inquiry* 57: 431–446.

Wolf, Charlotte. 1986. "Legitimation of Oppression: Response And Reflexivity." *Symbolic Interaction* 9: 217–234.

Zipter, Yvonne. 1988. *Diamonds Are A Dyke's Best Friend: Reflections, Reminiscences, And Reports from The Field on The Lesbian National Pastime.* Ithaca, NY: Firebrand Books.

33

Masculinity as Homophobia: Fear, Shame, and Silence in the Construction of Gender Identity

Michael S. Kimmel

We think of manhood as eternal, a timeless essence that resides deep in the heart of every man. We think of manhood as a thing, a quality that one either has or doesn't have. We think of manhood as innate, residing in the particular biological composition of the human male, the result of androgens or the possession of a penis. We think of manhood as a transcendent tangible property that each man must manifest in the world; the reward presented with great ceremony to a young novice by his elder for having successfully completed an arduous initiation ritual. In the words of poet Robert Bly (1990), "the structure at the bottom of the male psyche is still as firm as it was twenty thousand years ago" (230)....

This idea that manhood is socially constructed and historically shifting should not be understood as a loss, that something is being taken away from men. In fact, it gives us something extraordinarily valuable—agency, the capacity to act. It gives us a sense of historical possibilities to replace the despondent resignation that invariably attends timeless, ahistorical essentialisms. Our behaviors are not simply "just human nature," because "boys will be boys." From the materials we find around us in our culture—other people, ideas, objects—we actively create our worlds, our identities. Men, both individually and collectively, can change....

4 Sexism

Masculinity as a Homosocial Enactment

Other men: We are under the constant careful scrutiny of other men. Other men watch us, rank us, grant our acceptance into the realm of manhood. Manhood is demonstrated for other men's approval. It is other men who evaluate the performance. Literary critic David Leverenz (1991) argues that "ideologies of manhood have functioned primarily in relation to the gaze of male peers and male authority" (769). Think of how men boast to one another of their accomplishments—from their latest sexual conquest to the size of the fish they caught—and how we constantly parade the markers of manhood—wealth, power, status, sexy women—in front of other men, desperate for their approval.

That men prove their manhood in the eyes of other men is both a consequence of sexism and one of its chief props. "Women have, in men's minds, such a low place on the social ladder of this country that it's useless to define yourself in terms of a woman," noted playwright David Mamet.

"What men need is men's approval." Women become a kind of currency that men use to improve their ranking on the masculine social scale. (Even those moments of heroic conquest of women carry, I believe, a current of homosocial evaluation.) Masculinity is a *homosocial* enactment. We test ourselves, perform heroic feats, take enormous risks, all because we want other men to grant us our manhood.

Masculinity as a homosocial enactment is fraught with danger, with the risk of failure, and with intense relentless competition. "Every man you meet has a rating or an estimate of himself which he never loses or forgets," wrote Kenneth Wayne (1912) in his popular turn-of-the-century advice book. "A man has his own rating, and instantly he lays it alongside of the other man" (18). Almost a century later, another man remarked to psychologist Sam Osherson (1992) that "[b]y the time you're an adult, it's easy to think you're always in competition with men, for the attention of women, in sports, at work" (291). . . .

Homophobia is a central organizing principle of our cultural definition of manhood. Homophobia is more than the irrational fear of gay men, more than the fear that we might be perceived as gay. "The word 'faggot' has nothing to do with homosexual experience or even with fears of homosexuals," writes David Leverenz. "It comes out of the depths of manhood: a label of ultimate contempt for anyone who seems sissy, untough, uncool" (1986, 455). Homophobia is the fear that other men will unmask us, emasculate us, reveal to us and the world that we do not measure up, that we are not real men. We are afraid to let other men see that fear. Fear makes us ashamed, because the recognition of fear in ourselves is proof to ourselves that we are not as manly as we pretend, that we are, like the young man in a poem by Yeats, "one that ruffles in a manly pose for all his timid heart." Our fear is the fear of humiliation. We are ashamed to be afraid.

Shame leads to silence—the silence that keeps other people believing that we actually approve of the things that are done to women, to minorities, to gays and lesbians in our culture. The frightened silence as we scurry past a woman being hassled by men on the street. That furtive silence when men make sexist or racist jokes in a bar. That clammy-handed silence when guys in the office make gay-bashing jokes. Our fears are the sources of our silences, and men's silence is what keeps the system running. This might help to explain why women often complain that their male friends or partners are often so understanding when they are alone and yet laugh at sexist jokes or even make those jokes themselves when they are out with a group.

The fear of being seen as a sissy dominates the cultural definitions of manhood. It starts so early. "Boys among boys are ashamed to be unmanly," wrote one educator in 1871 (cited in Rotundo 1993, 264). I have a standing bet with a friend that I can walk onto any playground in America where 6-year-old boys are happily playing and by asking one question, I can provoke a fight. That question is simple: "Who's a sissy around here?" Once posed, the challenge is made. One of two things is likely to happen. One boy will accuse another of being a sissy, to which that boy will respond that he is not a sissy, that the first boy is. They may have to fight it out to see who's lying. Or a whole group of boys will surround one boy and all shout, "He is! He is!" That boy will either burst into tears and run home crying, disgraced, or he will have to take on several boys at once, to prove that he's not a sissy. (And what will his father or older brothers tell him if he chooses to run home crying?) It will be some time before he regains any sense of self-respect.

Violence is often the single most evident marker of manhood. Rather it is the willingness to fight, the desire to fight. The origin of our expression that one "has a chip on one's shoulder" lies in the practice of an adolescent boy in the country or small town at the turn of the century, who would literally walk around with a chip of wood balanced on his shoulder—a signal of his readiness to fight with anyone who would take the initiative of knocking the chip off (see Gorer 1964, 38; Mead 1965).

As adolescents, we learn that our peers are a kind of gender police, constantly threatening to unmask us as feminine, as sissies. One of the favorite tricks when I was an adolescent was to ask a boy to look at his fingernails. If he held his palm toward his face and curled his fingers back to see them, he passed the test. He'd looked at his nails "like a man." But if he held the back of his hand away from his face, and looked at his fingernails with arm outstretched, he was immediately ridiculed as a sissy.

As young men we are constantly riding those gender boundaries, checking the fences we have constructed on the perimeter, making sure that nothing even remotely feminine might show through. The possibilities of being unmasked are everywhere. Even the most seemingly insignificant thing can pose a threat or activate that haunting terror. On the day the students in my course "Sociology of Men and Masculinities" were scheduled to discuss homophobia and male-male friendships, one student provided a touching illustration. Noting that it was a beautiful day, the first day of spring after a brutal northeast winter, he decided to wear shorts to class. "I had this really nice pair of new Madras shorts," he commented. "But then I thought to myself, these shorts have lavender and pink in them. Today's class topic is homophobia. Maybe today is not the best day to wear these shorts."

Our efforts to maintain a manly front cover everything we do. What we wear. How we talk. How we walk. What we eat. Every mannerism, every movement contains a coded gender language. Think, for example, of how you would answer the question: How do you "know" if a man is homosexual? When I ask this question in classes or workshops, respondents invariably provide a pretty standard list of stereotypically effeminate behaviors. He walks a certain way, talks a certain way, acts a certain way. He's very emotional; he shows his feelings. One woman commented that she "knows" a man is gay if he really cares about her; another said she knows he's gay if he shows no interest in her, if he leaves her alone.

Now alter the question and imagine what heterosexual men do to make sure no one could possibly get the "wrong idea" about them. Responses typically refer to the original stereotypes, this time as a set of negative rules about behavior. Never dress that way. Never talk or walk that way. Never show your feelings or get emotional. Always be prepared to demonstrate sexual interest in women that you meet, so it is impossible for any woman to get the wrong idea about you. In this sense, homophobia, the fear of being perceived as gay, as not a real man, keeps men exaggerating all the traditional rules of masculinity, including sexual predation with women. Homophobia and sexism go hand in hand.

The stakes of perceived sissydom are enormous—sometimes matters of life and death. We take enormous risks to prove our manhood, exposing ourselves disproportionately to health risks, workplace hazards, and stress-related illnesses. Men commit suicide three times as often as women.... In one survey, women and men were asked what they were most afraid of. Women responded that they were most afraid of being raped and murdered. Men responded that they were most afraid of being laughed at (Noble 1992, 105–6).

Homophobia as a Cause of Sexism, Heterosexism, and Racism

Homophobia is intimately interwoven with both sexism and racism. The fear—sometimes conscious, sometimes not—that others might perceive us as homosexual propels men to enact all manner of exaggerated masculine behaviors and attitudes to make sure that no one could possibly get the wrong idea about us. One of the centerpieces of that exaggerated masculinity is putting women down, both by excluding them from the public sphere and by the quotidian put-downs in speech and behaviors that organize the daily life of the American man. Women and gay men become the "other" against which heterosexual men project their identities, against whom they stack the decks so as to compete in a situation in which they will always win, so that by suppressing them, men can stake a claim for their own manhood. Women threaten emasculation by representing the home, workplace, and familial responsibility, the negation of fun. Gay men have historically played the role of the consummate sissy in the American popular mind because homosexuality is seen as an inversion of normal gender development. There have been other "others." Through American history, various groups have represented the sissy, the non-men against whom American men played out their definitions of manhood, often with vicious results. In fact, these changing groups provide an interesting lesson in American historical development.

At the turn of the 19th century, it was Europeans and children who provided the contrast for American men. The "true American was vigorous, manly, and direct, not effete and corrupt like the supposed Europeans," writes Rupert Wilkinson (1986). "He was plain rather than ornamented, rugged rather than luxury seeking, a liberty loving common man or natural gentleman rather than an aristocratic oppressor or servile minion" (96). The "real man" of the early nineteenth century was neither noble nor serf. By the middle of the century, black slaves had replaced the effete nobleman. Slaves were seen as dependent, helpless men, incapable of defending their women and children, and therefore less than manly. Native Americans were cast as foolish and naive children, so they could be infantilized as the "Red Children of the Great White Father" and therefore excluded from full manhood.

By the end of the century, new European immigrants were also added to the list of the unreal men, especially the Irish and Italians, who were seen as too passionate and emotionally volatile to remain controlled sturdy oaks, and Jews, who were seen as too bookishly effete and too physically puny to truly measure up. In the mid-twentieth century, it was also Asians—first the Japanese during the Second World War, and more recently, the Vietnamese during the Vietnam War—who have served as unmanly templates against which American men have hurled their gendered rage. Asian men were seen as small, soft, and effeminate—hardly men at all.

Such a list of "hyphenated" Americans—Italian-, Jewish-, Irish-, African-, Native-, Asian-, gay—composes the majority of American men. So manhood is only possible for a distinct minority, and the definition has been constructed to prevent the others from achieving it. Interestingly, this emasculation of one's enemies has a flip side—and one that is equally gendered. These very groups that have historically been cast as less than manly were also, often simultaneously, cast as hypermasculine, as sexually aggressive, violent rapacious beasts, against whom "civilized" men must take a decisive stand and thereby rescue civilization. Thus black men were depicted as rampaging sexual beasts, women as carnivorously carnal, gay men as sexually insatiable, southern European men as sexually predatory and voracious, and Asian men as vicious and cruel torturers who were immorally disinterested in life itself, willing to sacrifice their entire people for their whims. But whether one saw these groups as effeminate sissies or as brutal uncivilized savages, the terms with which they were perceived were gendered. These groups become the "others," the screens against which traditional conceptions of manhood were developed.

Being seen as unmanly is a fear that propels American men to deny manhood to others, as a way of proving the unprovable—that one is fully manly. Masculinity becomes a defense against the perceived threat of humiliation in the eyes of other men, enacted through a "sequence of postures"—things we might say, or do, or even think, that, if we thought carefully about them, would make us ashamed of ourselves (Savran 1992, 16). After all, how many of us have made homophobic or sexist remarks, or told racist jokes, or made lewd comments to women on the street? How many of us have translated those ideas and those words into actions, by physically attacking gay men, or forcing or cajoling a woman to have sex even though she didn't really want to because it was important to score?

Power and Powerlessness in the Lives of Men

I have argued that homophobia, men's fear of other men, is the animating condition of the dominant definition of masculinity in America, that the reigning definition of masculinity is a defensive effort to prevent being emasculated. In our efforts to suppress or overcome those fears, the dominant culture exacts a tremendous price from those deemed less than fully manly: women, gay men, nonnative-born men, men of color. This perspective may help clarify a paradox in men's lives, a paradox in which men have virtually all the power and yet do not feel powerful (see Kaufman 1993).

Manhood is equated with power—over women, over other men. Everywhere we look, we see the institutional expression of that power—in state and national legislatures, on the boards of directors of every major U.S. corporation or law firm, and in every school and hospital administration. Women have long understood this, and feminist women have spent the past three decades challenging both the public and the private expressions of men's power and acknowledging their fear of men. Feminism as a set of theories both explains women's fear of men and empowers women to confront it both publicly and privately. Feminist women have theorized that masculinity is about the drive for domination, the drive for power, for conquest.

This feminist definition of masculinity as the drive for power is theorized from women's point of view. It is how women experience masculinity. But it assumes a symmetry between the public and the private that does not conform to men's experiences. Feminists observe that women, as a group, do not hold power in our society. They also observe that individually, they, as women, do not feel powerful. They feel afraid, vulnerable. Their observation of the social reality and their individual experiences are therefore symmetrical. Feminism also observes that men, as a group, are in power. Thus, with the same symmetry, feminism has tended to assume that individually men must feel powerful.

This is why the feminist critique of masculinity often falls on deaf ears with men. When confronted with the analysis that men have all the power, many men react incredulously. "What do you mean, men have all the power?" they ask. "What are you talking about? My wife bosses me around. My kids boss me around. My boss bosses me around. I have no power at all! I'm completely powerless!"

Men's feelings are not the feelings of the powerful, but of those who see themselves as powerless. These are the feelings that come inevitably from the discontinuity between the social and the psychological, between the aggregate analysis that reveals how men are in power as a group and the pyschological fact that they do not feel powerful as individuals. They are the feelings of men who were raised to believe themselves entitled to feel that power, but do not feel it. No wonder many men are frustrated and angry. This may explain the recent popularity of those workshops and retreats designed to help men to claim their "inner" power, their "deep manhood," or their "warrior within." . . .

The dimension of power is now reinserted into men's experience not only as the product of individual experience but also as the product of relations with other men. In this sense, men's experience of powerlessness is *real*—the men actually feel it and certainly act on it—but it is not *true*, that is, it does not accurately describe their condition. In contrast to women's lives, men's lives are structured around relationships of power and men's differential access to power, as well as the differential access to that power of men as a group. Our imperfect analysis of our own situation leads us to believe that we men need more power, rather than leading us to support feminists' efforts to rearrange power relationships along more equitable lines.

Philosopher Hannah Arendt (1970) fully understood this contradictory experience of social and individual power:

> Power corresponds to the human ability not just to act but to act in concert. Power is never the property of an individual; it belongs to a group and remains in existence only so long as the group keeps together. When we say of somebody that he is "in power" we actually refer to his being empowered by a certain number of people to act in their name. The moment the group, from which the power originated to begin with . . . disappears, "his power" also vanishes. (44)

Why, then, do American men feel so powerless? Part of the answer is because we've constructed the rules of manhood so that only the tiniest fraction of men come to believe that they are the biggest of wheels, the sturdiest of oaks, the most virulent repudiators of femininity, the most daring and aggressive. We've managed to disempower the overwhelming majority of American men by other means—such as discriminating on the basis of race, class, ethnicity, age, or sexual preference.

Masculinist retreats to retrieve deep, wounded, masculinity are but one of the ways in which American men currently struggle with their fears and their shame. Unfortunately, at the very moment that they work to break down the isolation that governs men's lives, as they enable men to express those fears and that shame, they ignore the social power that men continue to exert over women and the privileges from which they (as the middle-aged, middle-class white men who largely make up these retreats) continue to benefit—regardless of their experiences as wounded victims of oppressive male socialization.

Others still rehearse the politics of exclusion, as if by clearing away the playing field of secure gender identity of any that we deem less than manly—women, gay men, nonnative-born men, men of color—middle-class, straight, white men can reground their sense of themselves without those haunting fears and that deep shame that they are unmanly and will be exposed by other men. This is the manhood of racism, of sexism, of homophobia. It is the manhood that is so chronically insecure that it trembles at the idea of lifting the ban on gays in the military, that is so threatened by women in the workplace that women become the targets of sexual harassment, that is so deeply frightened of equality that it must ensure that the playing field of male competition remains stacked against all newcomers to the game.

Exclusion and escape have been the dominant methods American men have used to keep their fears of humiliation at bay. The fear of emasculation by other men, of being humiliated, of being seen as a sissy, is the leitmotif in my reading of the history of American manhood. Masculinity has become a relentless test by which we prove to other men, to women, and ultimately to ourselves, that we have successfully mastered the part. The restlessness that men feel today is nothing new in American history; we have been anxious and restless for almost two centuries. Neither exclusion nor escape has ever brought us the relief we've sought, and there is no reason to think that either will solve our problems now. Peace of mind, relief from gender struggle, will come only from a politics of inclusion, not exclusion, from standing up for equality and justice, and not by running away.

References

Arendt, H. (1970). *On Revolution.* New York: Viking.

Bly, R. (1990). *Iron John: A Book about Men.* Reading, Mass.: Addison-Wesley.

Gorer, G. (1964). *The American People: A Study in National Character.* New York: Norton.

Kaufman, M. (1993). *Cracking the Armour: Power and Pain in the Lives of Men.* Toronto: Viking Canada.

Leverenz, D. (1986). "Manhood, Humiliation and Public Life: Some Stories." *Southwest Review* 71, Fall.

Leverenz, D. (1991). "The Last Real Man in America: From Natty Bumppo to Batman." *American Literary Review* 3.

Mead, M. (1965). *And Keep Your Powder Dry.* New York: William Morrow.

Noble, V. (1992). "A Helping Hand from the Guys." In K. L. Hagan, ed., *Women Respond to the Men's Movement.* San Francisco: HarperCollins.

Osherson, S. (1992). *Wrestling with Love: How Men Struggle with Intimacy, with Women, Children, Parents, and Each Other.* New York: Fawcett.

Rotundo, E. A. (1993). *American Manhood: Transformations in Masculinity from the Revolution to the Modern Era.* New York: Basic Books.

Savran, D. (1992). *Communists, Cowboys and Queers: The Politics of Masculinity in the Work of Arthur Miller and Tennessee Williams.* Minneapolis: University of Minnesota Press.

Wayne, K. (1912). *Building the Young Man.* Chicago: A. C. McClurg.

Wilkinson, R. (1986). *American Tough: The Tough-Guy Tradition and American Character.* New York: Harper and Row.

SCENTS OF TIME • BATTLE OF THE BONES • NANOMEDICINE

THE SCIENCES

PUBLISHED BY THE NEW YORK ACADEMY OF SCIENCES • JULY/AUGUST 2000 • $3.95
CANADA $4.95

2000 NATIONAL MAGAZINE AWARD • BEST ESSAY

THE FIVE SEXES, REVISITED
The Varieties of Sex Will Test Medical Values and Social Norms

by ANNE FAUSTO-STERLING

ESSAYS & COMMENT

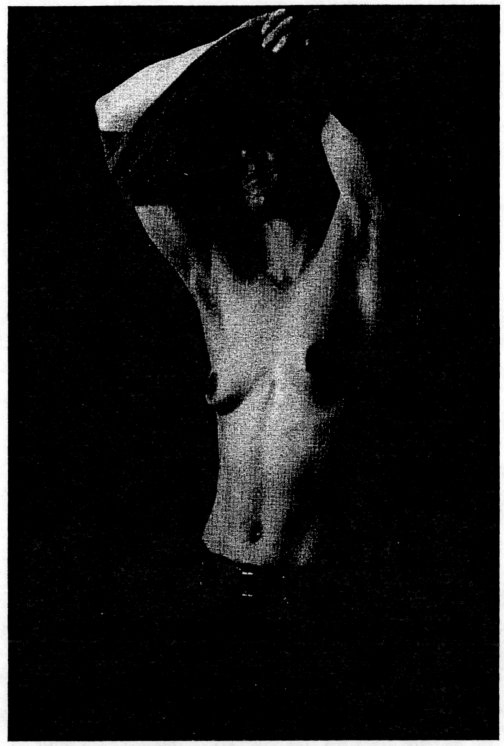

De LaGrace Volcano, Jack Unveiled, 1994

THE FIVE SEXES, REVISITED

The emerging recognition that people come in bewildering sexual varieties is testing medical values and social norms

BY ANNE FAUSTO-STERLING

As Cheryl Chase stepped to the front of the packed meeting room in the Sheraton Boston Hotel, nervous coughs made the tension audible. Chase, an activist for intersexual rights, had been invited to address the May 2000 meeting of the Lawson Wilkins Pediatric Endocrine Society (LWPES), the largest organization in the United States for specialists in children's hormones. Her talk would be the grand finale to a four-hour symposium on the treatment of genital ambiguity in newborns, infants born with a mixture of both male and female anatomy, or genitals that appear to differ from their chromosomal sex. The topic was hardly a novel one to the assembled physicians.

Yet Chase's appearance before the group was remarkable. Three and a half years earlier, the American Academy of Pediatrics had refused her request for a chance to present the patients' viewpoint on the treatment of genital ambiguity, dismissing Chase and her supporters as "zealots." About two dozen intersex people had responded by throwing up a picket line. The Intersex Society of North America (ISNA) even issued a press release: "Hermaphrodites Target Kiddie Docs."

It had done my 1960s street-activist heart good. In the short run, I said to Chase at the time, the picketing would make people angry. But eventually, I assured her, the doors then closed would open. Now, as Chase began to address the physicians at their own convention, that prediction was coming true. Her talk, titled "Sexual Ambiguity: The Patient-Centered Approach," was a measured critique of the near-universal practice of performing immediate, "corrective" surgery on thousands of infants born each year with ambiguous genitalia. Chase herself lives with the consequences of such surgery. Yet her audience, the very endocrinologists and surgeons Chase was accusing of reacting with "surgery and shame," received her with respect. Even more remarkably, many of the speakers who preceded her at the session had already spoken of the need to scrap current practices in favor of treatments more centered on psychological counseling.

What led to such a dramatic reversal of fortune? Certainly, Chase's talk at the LWPES symposium was a vindication of her persistence in seeking attention for her cause. But her invitation to speak was also a watershed in the evolving discussion about how to treat children with ambiguous genitalia. And that discussion, in turn, is the tip of a biocultural iceberg—the gender iceberg—that continues to rock both medicine and our culture at large.

Chase made her first national appearance in 1993, in these very pages, announcing the formation of ISNA in a letter responding to an essay I had written for *The Sciences*, titled "The Five Sexes" [March/April 1993]. In that article I argued that the two-sex system embedded in our society is not adequate to encompass the full spectrum of human sexuality. In its place, I suggested a five-sex system. In addition to males and females, I included "herms" (named after true hermaphrodites, people born with both a testis and an ovary); "merms" (male pseudohermaphrodites, who are born with testes and some aspect of female genitalia); and "ferms" (female pseudohermaphrodites, who have ovaries combined with some aspect of male genitalia).

I had intended to be provocative, but I had also written with tongue firmly in cheek. So I was surprised by the extent of the controversy the article unleashed. Right-wing Christians were outraged, and connected my idea of five sexes with the United Nations–sponsored Fourth World Conference on Women, held in Beijing in September 1995. At the same time, the article delighted others who felt constrained by the current sex and gender system.

Clearly, I had struck a nerve. The fact that so many people could get riled up by my proposal to revamp our sex and gender system suggested that change—as well as resistance to it—might be in the offing. Indeed, a lot has changed since 1993, and I like to think that my article was an important stimulus. As if from nowhere, intersexuals are materializing before our very eyes. Like Chase, many have become political organizers, who lobby physicians and politicians to change current treatment practices. But more generally, though perhaps no less provocatively, the boundaries separating masculine and feminine seem harder than ever to define.

Some find the changes under way deeply disturbing. Others find them liberating.

WHO IS AN INTERSEXUAL—AND HOW MANY intersexuals are there? The concept of intersexuality is rooted in the very ideas of male and female. In the idealized, Platonic, biological world, human beings are divided into two kinds: a perfect-

MUCH HAS CHANGED SINCE 1993.
Intersexuals have materialized
before our very eyes.

July/August 2000 • THE SCIENCES 19

ly dimorphic species. Males have an X and a Y chromosome, testes, a penis and all of the appropriate internal plumbing for delivering urine and semen to the outside world. They also have well-known secondary sexual characteristics, including a muscular build and facial hair. Women have two X chromosomes, ovaries, all of the internal plumbing to transport urine and ova to the outside world, a system to support pregnancy and fetal development, as well as a variety of recognizable secondary sexual characteristics.

That idealized story papers over many obvious caveats: some women have facial hair, some men have none; some women speak with deep voices, some men veritably squeak. Less well known is the fact that, on close inspection, absolute dimorphism disintegrates even at the level of basic biology. Chromosomes, hormones, the internal sex structures, the gonads and the external genitalia all vary more than most people realize. Those born outside of the Platonic dimorphic mold are called intersexuals.

In "The Five Sexes" I reported an estimate by a psychologist expert in the treatment of intersexuals, suggesting that some 4 percent of all live births are intersexual. Then, together with a group of Brown University undergraduates, I set out to conduct the first systematic assessment of the available data on intersexual birthrates. We scoured the medical literature for estimates of the frequency of various categories of intersexuality, from additional chromosomes to mixed gonads, hormones and genitalia. For some conditions we could find only anecdotal evidence; for most, however, numbers exist. On the basis of that evidence, we calculated that for every 1,000 children born, seventeen are intersexual in some form. That number—1.7 percent—is a ballpark estimate, not a precise count, though we believe it is more accurate than the 4 percent I reported.

Our figure represents all chromosomal, anatomical and hormonal exceptions to the dimorphic ideal; the number of intersexuals who might, potentially, be subject to surgery as infants is smaller—probably between one in 1,000 and one in 2,000 live births. Furthermore, because some populations possess the relevant genes at high frequency, the intersexual birthrate is not uniform throughout the world.

Consider, for instance, the gene for congenital adrenal hyperplasia (CAH). When the CAH gene is inherited from both parents, it leads to a baby with masculinized external genitalia who possesses two X chromosomes and the internal reproductive organs of a potentially fertile woman. The frequency of the gene varies widely around the world: in New Zealand it occurs in only forty-three children per million; among the Yupik Eskimo of southwestern Alaska, its frequency is 3,500 per million.

INTERSEXUALITY HAS ALWAYS BEEN TO SOME extent a matter of definition. And in the past century physicians have been the ones who defined children as intersexual—and provided the remedies. When only the chromosomes are unusual, but the external genitalia and gonads clearly indicate either a male or a female, physicians do not advocate intervention. Indeed, it is not clear what kind of intervention could be advocated in such cases. But the story is quite different when infants are born with mixed genitalia, or with external genitals that seem at odds with the baby's gonads.

Most clinics now specializing in the treatment of intersex babies rely on case-management principles developed in the 1950s by the psychologist John Money and the psychiatrists Joan G. Hampson and John L. Hampson, all of Johns Hopkins University in Baltimore, Maryland. Money believed that gender identity is completely malleable for about eighteen months after birth. Thus, he argued, when a treatment team is presented with an infant who has ambiguous genitalia, the team could make a gender assignment solely on the basis of what made the best surgical sense. The physicians could then simply encourage the parents to raise the child according to

the surgically assigned gender. Following that course, most physicians maintained, would eliminate psychological distress for both the patient and the parents. Indeed, treatment teams were never to use such words as "intersex" or "hermaphrodite"; instead, they were to tell parents that nature intended the baby to be the boy or the girl that the physicians had determined it was. Through surgery, the physicians were merely completing nature's intention.

Although Money and the Hampsons published detailed case studies of intersex children who they said had adjusted well to their gender assignments, Money thought one case in particular proved his theory. It was a dramatic example, inasmuch as it did not involve intersexuality at all: one of a pair of identical twin boys lost his penis as a result of a circumcision accident. Money recommended that "John" (as he came to be known in a later case study) be surgically turned into "Joan" and raised as a girl. In time, Joan grew to love wearing dresses and having her hair done. Money proudly proclaimed the sex reassignment a success.

But as recently chronicled by John Colapinto, in his book *As Nature Made Him,* Joan—now known to be an adult male named David Reimer—eventually rejected his female assignment. Even without a functioning penis and testes (which had been removed as part of the reassignment) John/Joan sought masculinizing medication, and married a woman with children (whom he adopted).

Since the full conclusion to the John/Joan story came to light, other individuals who were reassigned as males or females shortly after birth but who later rejected their early assignments have come forward. So, too, have cases in which the reassignment has worked—at least into the subject's mid-twenties. But even then the aftermath of the surgery can be problematic. Genital surgery often leaves scars that reduce sexual sensitivity. Chase herself had a complete clitoridectomy, a procedure that is less frequently performed on intersexuals today. But the newer surgeries, which reduce the size of the clitoral shaft, still greatly reduce sensitivity.

THE REVELATION OF CASES OF FAILED REASSIGN-ments and the emergence of intersex activism have led an increasing number of pediatric endocrinologists, urologists and psychologists to reexamine the wisdom of early genital surgery. For example, in a talk that preceded Chase's at the LWPES meeting, the medical ethicist Laurence B. McCullough of the Center for Medical Ethics and Health Policy at Baylor College of Medicine in Houston, Texas, introduced an ethical framework for the treatment of children with ambiguous genitalia. Because sex phenotype (the manifestation of genetically and embryologically determined sexual characteristics) and gender presentation (the sex role projected by the individual in society) are highly variable, McCullough argues, the various forms of intersexuality should be defined as normal. All of them fall within the statistically expected variability of sex and gender. Furthermore, though certain disease states may accompany some forms of intersexuality, and may require medical intervention, intersexual conditions are not themselves diseases.

McCullough also contends that in the process of assigning gender, physicians should minimize what he calls irreversible assignments: taking steps such as the surgical removal or modification of gonads or genitalia that the patient may one day want to have reversed. Finally, McCullough urges physicians to abandon their practice of treating the birth of a child with genital ambiguity as a medical or social emergency. Instead, they should take the

Janine Antoni, Mom and Dad, *1994*

time to perform a thorough medical workup and should disclose everything to the parents, including the uncertainties about the final outcome. The treatment mantra, in other words, should be therapy, not surgery.

I believe a new treatment protocol for intersex infants, similar to the one outlined by McCullough, is close at hand. Treatment should combine some basic medical and ethical principles with a practical but less drastic approach to the birth of a mixed-sex child. As a first step, surgery on infants should be performed only to save the child's life or to substantially improve the child's physical well-being. Physicians may assign a sex—male or female—to an intersex infant on the basis of the probability that the child's particular condition will lead to the formation of a particular gender identity. At the same time, though, practitioners ought to be humble enough to recognize that as the child grows, he or she may reject the assignment—and they should be wise enough to listen to what the child has to say. Most important, parents should have access to the full range of information and options available to them.

Sex assignments made shortly after birth are only the

beginning of a long journey. Consider, for instance, the life of Max Beck: Born intersexual, Max was surgically assigned as a female and consistently raised as such. Had her medical team followed her into her early twenties, they would have deemed her assignment a success because she was married to a man. (It should be noted that success in gender assignment has traditionally been defined as living in that gender as a heterosexual.) Within a few years, however, Beck had come out as a butch lesbian; now in her mid-thirties, Beck has become a man and married his lesbian partner, who (through the miracles of modern reproductive technology) recently gave birth to a girl.

Transsexuals, people who have an emotional gender at odds with their physical sex, once described themselves in terms of dimorphic absolutes—males trapped in female bodies, or vice versa. As such, they sought psychological relief through surgery. Although many still do, some so-called transgendered people today are content to inhabit a more ambiguous zone. A male-to-female transsexual, for instance, may come out as a lesbian. Jane, born a physiological male, is now in her late thirties and living with her wife, whom she married when her name was still John. Jane takes hormones to feminize herself, but they have not yet interfered with her ability to engage in intercourse as a man. In her mind Jane has a lesbian relationship with her wife, though she views their intimate moments as a cross between lesbian and heterosexual sex.

It might seem natural to regard intersexuals and transgendered people as living midway between the poles of male and female. But male and female, masculine and feminine, cannot be parsed as some kind of continuum. Rather, sex and gender are best conceptualized as points in a multidimensional space. For some time, experts on gender development have distinguished between sex at the genetic level and at the cellular level (sex-specific gene expression, X and Y chromosomes); at the hormonal level (in the fetus, during childhood and after puberty); and at the anatomical level (genitals and secondary sexual characteristics). Gender identity presumably emerges from all of those corporeal aspects via some poorly understood interaction with environment and experience. What has become increasingly clear is that one can find levels of masculinity and femininity in almost every possible permutation. A chromosomal, hormonal and genital male (or female) may emerge with a female (or male) gender identity. Or a chromosomal female with male fetal hormones and masculinized genitalia—but with female pubertal hormones—may develop a female gender identity.

THE MEDICAL AND SCIENTIFIC COMMUNITIES have yet to adopt a language that is capable of describing such diversity. In her book *Hermaphrodites and the Medical Invention of Sex*, the historian and medical ethicist Alice Domurat Dreger of Michigan State University in East Lansing documents the emergence of current medical systems for classifying gender ambiguity.

A PERSON WHO PROJECTS *a social gender at odds with his or her genitals may die for the transgression.*

The current usage remains rooted in the Victorian approach to sex. The logical structure of the commonly used terms "true hermaphrodite," "male pseudohermaphrodite" and "female pseudohermaphrodite" indicates that only the so-called true hermaphrodite is a genuine mix of male and female. The others, no matter how confusing their body parts, are really hidden males or females. Because true hermaphrodites are rare—possibly only one in 100,000—such a classification system supports the idea that human beings are an absolutely dimorphic species.

At the dawn of the twenty-first century, when the variability of gender seems so visible, such a position is hard to maintain. And here, too, the old medical consensus has begun to crumble. Last fall the pediatric urologist Ian A. Aaronson of the Medical University of South Carolina in Charleston organized the North American Task Force on Intersexuality (NATFI) to review the clinical responses to genital ambiguity in infants. Key medical associations, such as the American Academy of Pediatrics, have endorsed NATFI. Specialists in surgery, endocrinology, psychology, ethics, psychiatry, genetics and public health, as well as intersex patient-advocate groups, have joined its ranks.

One of the goals of NATFI is to establish a new sex nomenclature. One proposal under consideration replaces the current system with emotionally neutral terminology that emphasizes developmental processes rather than preconceived gender categories. For example, Type I intersexes develop out of anomalous virilizing influences; Type II result from some interruption of virilization; and in Type III intersexes the gonads themselves may not have developed in the expected fashion.

WHAT IS CLEAR IS THAT SINCE 1993, modern society has moved beyond five sexes to a recognition that gender variation is normal and, for some people, an arena for playful exploration. Discussing my "five sexes" proposal in her book *Lessons from the Intersexed*, the psychologist Suzanne J. Kessler of the State University of New York at Purchase drives this point home with great effect:

The limitation with Fausto-Sterling's proposal is that . . . [it] still gives genitals . . . primary signifying status and ignores the fact that in the everyday world gender attributions are made without access to genital inspection. . . . What has primacy in everyday life is the gender that is performed, regardless of the flesh's configuration under the clothes.

I now agree with Kessler's assessment. It would be better for intersexuals and their supporters to turn everyone's focus away from genitals. Instead, as she suggests, one should acknowledge that people come in an even wider assortment of sexual identities and characteristics than mere genitals can distinguish. Some women may have "large clitorises or fused labia," whereas some men may have "small penises or misshapen scrota," as Kessler puts it, "phenotypes with no particular clinical or identity meaning."

22 THE SCIENCES · *July/August 2000*

Arthur Tress, Hermaphrodite, 1973

As clearheaded as Kessler's program is—and despite the progress made in the 1990s—our society is still far from that ideal. The intersexual or transgendered person who projects a social gender—what Kessler calls "cultural genitals"—that conflicts with his or her physical genitals still may die for the transgression. Hence legal protection for people whose cultural and physical genitals do not match is needed during the current transition to a more gender-diverse world. One easy step would be to eliminate the category of "gender" from official documents, such as driver's licenses and passports. Surely attributes both more visible (such as height, build and eye color) and less visible (fingerprints and genetic profiles) would be more expedient.

A more far-ranging agenda is presented in the International Bill of Gender Rights, adopted in 1995 at the fourth annual International Conference on Transgender Law and Employment Policy in Houston, Texas. It lists ten "gender rights," including the right to define one's own gender, the right to change one's physical gender if one so chooses and the right to marry whomever one wishes. The legal bases for such rights are being hammered out in the courts as I write and, most recently, through the establishment, in the state of Vermont, of legal same-sex domestic partnerships.

N<small>O ONE COULD HAVE FORESEEN SUCH</small> changes in 1993. And the idea that I played some role, however small, in reducing the pressure—from the medical community as well as from society at large—to flatten the diversity of human sexes into two diametrically opposed camps gives me pleasure.

Sometimes people suggest to me, with not a little horror, that I am arguing for a pastel world in which androgyny reigns and men and women are boringly the same. In my vision, however, strong colors coexist with pastels. There are and will continue to be highly masculine people out there; it's just that some of them are women. And some of the most feminine people I know happen to be men. •

A<small>NNE</small> F<small>AUSTO-</small>S<small>TERLING</small> *is a professor of biology and women's studies at Brown University. Portions of this article were adapted from her recent book* S<small>EXING THE</small> B<small>ODY</small> *(Basic Books, 2000).*

Am I Man Enough Yet? A Comparison of the Body Transition, Self-Labeling, and Sexual Orientation of Two Cohorts of Female-to-Male Transsexuals

Adam F. Yerke
Valory Mitchell

ABSTRACT. This study compares 2 cohorts of female-to-male (FtM) transsexuals, one who transitioned between 1969 and 1987 and the other, between 2000 and 2006. Eight individuals from each group were interviewed. Results revealed that FtMs transitioning since 2000 less often seek bottom surgery, use a greater variety of terms to label their gender identity, and more often identify themselves as attracted to both men and women (and, for some, being attracted to transgender persons). Findings suggest that the experience of feeling "man enough" (achieving satisfaction with his sex and/or gender) has changed for FtMs.

KEYWORDS. Transgender, transsexual, gender dysphoria, gender identity, sex reassignment surgery

An FtM, or female-to-male transsexual, or transgender man, is an individual who is born female and later lives as a man (Israel & Tarver, 1997). Generally, the goal of transition for FtMs is to experience greater satisfaction with their gender identity. FtMs reach fulfillment (and, it is hoped, reduce gender dysphoria, if present) by making transition choices that suit their individual sense of who they are (and who they aren't). As individuals move through transition, FtMs may ask themselves, "Am I man[1] enough yet?" Making changes to one's body, choosing labels to identify one's gender, and identifying one's sexual orientation are all developmental tasks that many FtMs go through while answering this question. This study compares two cohorts of FtMs, one who transitioned between 1969 and 1987 and the other who transitioned between 2000 and 2006, on these three domains of decision making. Cross-cohort similarities and differences are described.

In the nearly 60 years since the first U.S. FtM's transition was reported (Devor, 1997), advances in social attitudes, medical technology, and in the availability of information, support, and medical care have led to substantial differences in the transition experience (Yerke, 2008). A half-century ago, Harry Benjamin, a physician and key figure in promoting awareness of transsexuality and its treatment, believed that gender dysphoric patients should be provided with what they have always desired: to live as the sex and/or gender with which they identify (Califia, 1997). Therefore, Benjamin asserted, sex reassignment surgery was the most appropriate treatment of gender dysphoria. Johns

---

Adam F. Yerke and Valory Mitchell are affiliated with the California School of Professional Psychology at Alliant International University in San Francisco, California.

Address correspondence to Valory Mitchell, California School of Professional Psychology at Alliant University, 1 Beach Street, San Francisco, CA 94133. E-mail: vmitchell@alliant.edu

Hopkins University Hospital concurred and became the first U.S. hospital to officially support sex-change operations in 1966 (Pauly & Edgerton, 1986). Following this well-regarded institution's example, by 1979, 20 major medical centers across the nation offered counseling, hormonal regimens, and operations (Ettner, 1999). The era of the gender clinics, with their strict standards and gatekeeping protocols, has gradually evolved, so that today the physical changes of transition can take place in a great variety of medical settings, including in community-based health care centers or with the assistance of private practitioners.

Symptoms of gender dysphoria must be present for a mental health professional to diagnose someone with gender identity disorder (American Psychiatric Association, 2000; Lev, 2004). Obtaining this diagnosis is usually a prerequisite for a transgender individual to receive hormonal or surgical options. Because of this requirement, both of the cohorts in this study must have asserted considerable gender dysphoria.

Decision-Making Domain #1: Physical Transition

As a part of his physical transition, an FtM may (or may not) undergo hormone therapy, chest surgery, and/or reproductive and/or genital surgery (Ettner, 1999). Several possibilities exist within each of these domains.

FtMs who undergo hormone treatment receive testosterone through an intramuscular injection, transdermal patch (Ettner, 1999), or topical cream or gel (Rachlin, Green, & Lombardi, 2008). As a result, masculinization occurs: deepening of the voice, clitoral enlargement, mild breast atrophy, increased facial and body hair, and male-pattern baldness (Ettner, 1999; Gooren, 1999; Meyer et al., 2001; Rachlin, 1999). The aim of "hormonal reassignment" is to reduce the secondary sex characteristics of a person's biological sex and to induce those of the preferred sex (Gooren, 1999). Hormone therapy helps FtMs feel and appear more like biological men (Meyer et al., 2001).

For many FtMs, chest surgery is an important component of transition, because the chest area is always visible, and the presence of breasts contradicts the visual cue that indicates that a person is presenting as a man (Ettner, 1999). In order to remove breast tissue, FtMs most often have some combination of liposuction, mastectomy, and/or chest reconstruction (Rachlin, 1999).

FtMs may decide on surgeries that remove their reproductive organs, such as hysterectomy, oophorectomy, and vaginectomy (Ettner, 1999; Meyer et al., 2001; Rachlin, 1999).

Two surgeries seek to modify the genitals so that they appear more like those of biological males. Phalloplasty attempts to construct a full-size phallus (Rachlin, 1999), providing the ability to engage in sexual intercourse and to urinate standing (Ettner, 1999). Most phalloplasty procedures take tissue from other body parts, such as the forearm, abdomen, or leg. The results of this surgery are inconsistent and imprecise; the procedures are costly and often involve complications (Tsoi, Kok, Yeo, & Ratnam, 1995). Metoidioplasty, an alternative genital reconstructive surgery (Ettner, 1999; Rachlin, 1999), releases the clitoris, which has been enlarged by hormones, from the clitoral hood and transforms it into a "microphallus." The urethra may be extended to allow FtMs to urinate standing (Kotula, 2002; Rachlin, 1999); however, the microphallus is not usually large enough to be used for intercourse (Ettner, 1999; Kotula, 2002; Rachlin, 1999). In addition, a scrotum may be created from the labia majora and filled with testicular implants (Kotula, 2002; Rachlin, 1999).

Decision-Making Domain #2: Self-Labeling

The language used to describe the transgender person varies depending on the source (Lev, 2004). The majority of transgender labels were created by those observing and treating this population, for example, researchers, scientists, physicians, and psychologists (Cromwell, 1999; Meyerowitz, 2002). Some terms are derived from derogatory names used against transgender persons (Feinberg, 1999). Among the interviewees, some, using the same term, are referring to different concepts, while the same concept may be defined differently among participants. In addition, new ways to identify oneself continue to emerge within the transgender community (Feinberg, 1998; Stryker & Whittle, 2006). Some people do not identify with any particular existing term and formulate their own ways of identification (Cromwell, 1999).

Sex and gender are usually presented dichotomously; that is, a person is either male or female, man or woman (Lev, 2004). Nontransgender (cisgender) persons usually label their sex and/or gender according to the sex they were assigned at birth (Etaugh & Bridges, 2004). From this perspective—a perspective shared by many of the early gender clinics of the 1960s and 70s—a successful transition would be indicated by an FtM identifying solely as male/man (Cromwell, 1999; Green, 2000). Sex and gender identity, however, may be more complex than this dichotomous theory suggests (Bockting, 2008; Valentine, 2006).

Transsexual

The term *transsexual* was first used by a psychiatrist to describe a woman who desired to be a man (Lothstein, 1983; Meyerowitz, 2002). This term continues to describe people who experience a persistent desire to live as the other sex and, as a result, embark upon a period of transition in which they live full time as the sex with which they identify (Brown & Rounsley, 1996). The term transsexual has also been applied to people who have physically altered their body through hormones or surgery, thus changing their sex (Cromwell, 1999).

Transgender

The term *transgender* is an umbrella term used to describe a variety of gender-variant people, including transsexuals, cross-dressers, and genderqueer individuals (Brown & Rounsley, 1996; Israel & Tarver, 1997). Individuals meeting the definition of a "transsexual" may also identify with this term.

Female-to-Male/FtM

The term *female-to-male* (FtM) refers to a person who was born female and currently identifies as a man (Brown & Rounsley, 1996; Israel & Tarver, 1997). This term may be used to describe a transgender person who is living temporarily, part time, or full time as the other gender (Israel & Tarver, 1997).

Genderqueer

Genderqueer (or androgynous) individuals do not identify with one gender or the other (Israel & Tarver, 1997). Androgynous people usually retain some characteristics of their biological sex and also adopt attributes of the other gender. Often, they do not want to be referred to as either a man or woman, since they do not identify with either gender.

Other terms that may be used by FtMs include *transperson, transman, F2M, or MtM* (*male-to-men*; Cromwell, 1999).

Decision-Making Domain #3: Sexual Orientation

Sexual orientation describes people's gendered sexual attractions: to people of the same sex (homosexual), the other sex (heterosexual), both sexes (bisexual), or neither (asexual; Israel & Tarver, 1997). Initially, transsexuals were thought to have little, if any, sexual interest (no matter the partner's gender), due to the disgust they felt for their bodies (Benjamin, 1966). Some believed that transsexuals were actually shame-filled homosexuals and that by transitioning, they would escape any homosexuality-related shame (Socarides, 1970).

In the 1970s and the early 80s, sexuality was taken into account when transsexuals were assessed by gender clinics for sex reassignment surgery (Meyerowitz, 2002). In order to receive treatment, an FtM needed to convince others that he would live as a "normal" man, including a persistent attraction to (and sexual behaviors with) nonhomosexual females (Lothstein, 1983). For some FtMs, this meant denying their true sexual feelings in order to transition (Bolin, 1988). More recent literature indicates that a person's sex and gender are not related to his or her sexual orientation (Bockting, Benner, & Coleman, 2009; Coleman, Bockting, & Gooren, 1993; Cromwell, 1999; Lev, 2004; Meyerowitz, 2002).

Therefore, FtMs may identify as heterosexual, homosexual, bisexual, or asexual (Meyerowitz, 2002). In addition, FtMs may be attracted to other transgender persons (Cromwell, 1999).

METHOD

Participants

Two groups, each with eight participants, were interviewed. Individuals were all born female-bodied and transitioned to living full time as men in all realms of life. Eight FtMs who transitioned between 2000 and 2006 made up the first group; the second group included eight FtMs who transitioned between 1969 and 1987. All participants had transitioned between age 18 and 30.

In order to recruit participants, flyers were distributed at gay, lesbian, bisexual, and transgender (GLBT) community centers, FtM support groups and conferences, and medical and psychological clinics that work with transgender clients. Flyers were also posted electronically in FtM discussion forums, online community pages, personal websites, and on craigslist. Participant-driven snowball sampling, through which each contact was invited to refer other potential participants, was also used.

Procedures

Potential participants contacted the interviewer by phone or e-mail, and interview appointments were set with qualified participants. In order to be included in the study, individuals were required to be born female-bodied and transitioned to living full time as men, in all realms of life. FtMs in the first of two groups needed to have transitioned within 5 years of the interview date, but not more recently than 6 months. This group was restricted to participants who were between the ages of 18 and 30 years; this age requirement existed so that only adults were included and so that the age at which both groups transitioned would be similar. Participants in the second group were included if they had transitioned at least 20 years prior to the interview date; no age restriction was set for Group 2. Interviews (60–90 min) were held either in person or by phone. The interview began with demographic questions, followed by a series of open-ended questions about the interviewee's transition process. Interviews were audiotaped.

Analysis

A two-to-three-page summary was written from the transcripts for each interview. Then, issue-focused analysis was used to identify common core themes and significant differences. These became coding categories. Returning to the interview transcripts, codes were attached to words, phrases, sentences, and whole paragraphs that signified a specific theme or concept (Miles & Huberman, 1994). The purpose of coding was to find a connection between the researcher's questions and the responses provided by the participant (Weiss, 1994). When coding the transcripts, the researcher also identified particular segments that served as illustrative quotes.

The coded segments were grouped into meaningful categories (Weiss, 1994). This sorting process entailed clustering codes together. In the next step, local integration, the researcher summarized the data in each category (Weiss, 1994) to allow the themes to be better identified and more condensed. The final stage of analysis, inclusive integration, consisted of the researcher compiling and integrating the themes that were found (Weiss, 1994). In this stage, all the identified categories were integrated with the research questions of the study.

When the above sequence of analysis was completed within each group, the two groups were compared. First, demographic information was compared across groups. Then, the researcher identified themes that were similar across groups, as well as themes that were unique to a group.

RESULTS

The study's main research question is whether FtMs' decisions and preferences about various aspects of physical transformation, self-labeling, and sexual orientation have changed over time and considers some explanations for these different choices. The Recent Transition Group consisted of participants who had transitioned between 2000 and 2006. Members of this group were between 21 and 30 years ($M = 25.88$). Seven were Caucasian, and one was Latino. Two were high school graduates; five held a bachelor's degree, including two currently in master's programs and one in a doctoral program; and one held a master's degree. One was married, four had a primary partner, and three were single.

The Early Transition Group had transitioned between 1969 and 1987. Members of this group were between 39 and 57 years ($M = 46.88$). Six were Caucasian, one was half Caucasian and half Latino, and one was half Caucasian and half American Indian. One was a high school graduate; one had a bachelor's degree; five held a master's degree, including one currently in a doctorate program; and one held a doctoral degree. Five were married, one had a primary partner, and two were single.

Physical/Medical Changes

While nearly all members of both cohorts elected to undertake hormone therapy and top surgery, they differed markedly by cohort on the decision to remove internal reproductive organs or experience genital reconstruction (see Table 1).

Hormones

All Recent Transition Group and Early Transition Group participants had hormone therapy as a part of their transition.

A Recent Transition Group participant explained,

> I was very emotional the night before. I was wondering, you know, is my ass gonna get really hairy, is my back gonna get really hairy, am I gonna look really really butch. I mean, can I still be like a pretty boy?—because I like being pretty, but I wanna be a pretty *boy* ... and I remember after I got my first shot I was just crying, like tears of happiness. It was like the best thing, and I documented my entire transition.

TABLE 1. Physical/Medical Changes for FtMs' Transitioning ($N = 16$)

	Recent Transition Group ($n = 8$)	Early Transition Group ($n = 8$)
Changes made		
Hormone treatment	8	8
Chest/Top surgery	7	8
Bottom surgery	0	8
Hysterectomy	0	7
Phalloplasty	0	4
Metoidioplasty	0	1
Changes desired for future		
Bottom surgery/ revision	3	4
Top surgery/ revision	2	0
Stop hormones to get pregnant	1	0

An Early Transition Group Participant said,

> Starting on hormones and having my voice change was just the greatest thing, because suddenly I could talk. Because previously I had not talked, under most conditions, because it would give me away. When I was around people I didn't know, I would always try to pass, so that meant being very unsociable because I was using a male name, so that was a huge change.

Top Surgery

Seven of eight participants from the Recent Transition Group had had top surgery, and all eight participants from the Early Transition Group had had top surgery. All participants who underwent top surgery reported having surgery after starting hormones. The one remaining participant wanted chest surgery, but was unable to have it because of his financial limitations and health complications.

A Recent Transition Group participant explained,

> I had chest surgery. That was a liberating thing, like the best thing I've ever done...I really didn't have much of a chest to start, but whatever was there, it did not feel

right...It was a really good experience. My mom was there to support. She went with me. She waited in the room and she helped me during with my drains and things like that so it was really good.

An Early Transition Group Participant said,

I guess for me that was a real important one, because it was so external and part of what I had really dreaded about my body ... I was pretty happy.

Bottom Surgery

No participants from the Recent Transition Group had had any type of bottom surgery. In contrast, all participants from the Early Transition Group had had some form of bottom surgery. Seven had had hysterectomies; four, phalloplasties; and one, a metoidioplasty. Six of the eight participants in this group had had chest surgery prior to any bottom surgeries.

A Recent Transition Group participant explained,

I've never felt very intensely that I should have had a penis at birth or like that belongs on me, so I'm not interested in an expensive and painful and not very effective genital surgery.

In contrast, Early Transition Group members were strongly committed to reproductive and genital surgeries, or felt they were necessary in order to legitimate the transition:

[The phalloplasty results] are great. I love making love and I've been real happy with that part of my life for 30 years. So, there's nothing I would change.

[The hysterectomy] was pretty easy. We had to just spend a big bunch of money and get a surgery done just to get a piece of paper to get married ... but it was not a priority.

When I was 22, I had a hysterectomy. I wanted that done really bad, so I actually had insurance pay for it.

[After the phalloplasty], there were three months I really couldn't go out ... That was kind of a lonely time because I really couldn't go anywhere. Friends came over, but I felt medically fragile.

I had a metoidioplasty done. So, I am as masculine as any other person. I have been in locker rooms and nobody has taken a second look at the size of what I have, so it's cool.

Future Changes?

Participants were asked if there were things not yet accomplished that they hoped to eventually include in their physical transition. The participant from the Recent Transition Group who had not had top surgery wanted it, and another participant wanted additional top surgery. No one from the Early Transition Group desired to have additional top surgery.

One participant from the Recent Transition Group wanted to discontinue hormones in order to get pregnant; this was not mentioned by any Early Transition Group participants.

None of the Recent Transition Group had had bottom surgery, and only three of eight participants reported interest in it in the future; none of these participants had any proximate plan to do so. Four of the eight had no interest in bottom surgery, and a fifth did not mention surgeries when asked about possible future changes. Participants' reasons included liking their current body, wanting to wait to see what surgical advances might develop, the high cost, and poor results.

Although all of the Early Transition Group members had had some form of bottom surgery, four of the eight participants from the Early Transition Group expressed interest in future bottom surgery; two of these would be revisions to previous bottom surgeries. One participant from the Early Transition Group had no plans for bottom surgery because he was aware of reports of dissatisfaction with the outcome.

Examples of comments from those not seeking bottom surgery were, from the Recent Transition Group:

> Bottom surgery is not really an option for me... It's really expensive for really crappy results... If the surgery was better... then it would definitely be an option for me.

> I think kind of casually about the concept of bottom surgery, but none of the options look that great to me, so I'm pretty much content at this point to wait and see what else develops or technologies get better or anything like that.

and from the Early Transition Group:

> I would like to get lower surgery, but I don't like any of the surgeries, until they can come up with a good urinary hookup... I would like to get lower surgery only if I can be guaranteed through one or more surgeries that I could stand up and take a leak.

Examples of comments from those planning bottom surgery expressing other themes were, from the Recent Transition Group:

> Certainly bottom surgery has crossed my mind. I don't know. I mean I would love it. Whenever I think about myself, fantasize about myself, I do have a penis, but reality wise, I just don't think it's there yet... I've really hated being a guinea pig and I feel like I've done that in therapy and interim with these hormones... So, if I ever were to have bottom surgery, I would want it to be exceptional, especially for the thousands of dollars.

> I would like to get bottom surgery, but technology doesn't have much to my liking. Phalloplasty is not something I will do. I don't feel that I would feel comfortable with the results and so that's not an option. Now, metoidioplasty procedure, something like that would be more realistic for me... I also need to get a hysterectomy, but that's in the plan...

and from the Early Transition Group:

> In '97, I did go back and try to have this urinary hookup done, which included a vaginectomy. [The doctor] never finished it. [He had to retire, due to medical problems] and I have yet to be able to find a physician who can finish it, but I'm getting closer. It's kind of like remodeling a house where people have come in and done stuff and done stuff and done stuff and people know how to do it from scratch, but they don't know how to like, fix what other people have done. So, that's been obviously not a positive piece of it.

> Yes... I'm getting the urethral hookup in the summer that I'm looking forward to.

> I think the only thing I'm seriously considering at this point is the clitoral free up and possibly the metoidioplasty... I would never get a urinary hookup because I've never seen one of those come out well, and you almost always end up with a fistula.

Choice of Gender Identity Terms

Both groups reported using several gender identity terms, and people in both groups used different identity terms in different contexts (e.g., identifying as FtM among other transgender people, but as a male among coworkers).

Those who had transitioned since 2000 (Recent Transition Group) used a slightly greater variety of gender identity terms. The Recent Transition Group used the following labels: man or male (six participants), female-to-male or FtM (five participants), transman (three participants), trans (one participant), feminine male (one participant), and genderqueer FtM (one participant).

The Early Transition Group used the following labels: man or male (eight participants), FtM (four participants), transman (two participants), transsexual (two participants), and transgender (one participant). While all of the Early Transition Group participants self-labeled as men, only three fourths of the Recent Transition Group did so. About half of the members of both groups self-labeled as FtM (four and five), and about one fourth of each group identified as transmen (two and three). Only those in the Early Transition Group labeled themselves as transsexual (two) or transgender (one), while only those in the Recent Transition Group labeled themselves as trans (one), feminine male (one) or genderqueer (one).

Statements from three participants in the Recent Transition Group provide examples of how these men think about the terms they use to describe themselves:

> I do have some definite pride around the FtM transman label. I think it is important for community building that we all try to identify ourselves and join together, so I do. I do very much identify with that label and then there's also parts of me that are very androgynous and maybe gender variant ... But certainly FtM transman describes me pretty well.

> I identify as FtM or trans ... or just male. I guess I identify as all three. I don't specially say, "Well, I'm FtM." I mean, most times I just say I'm trans, and I guess leave it at that.

> In regular day-to-day life, I just identify as male, but if I were to sit down and have a conversation about my gender I would be like, "Yeah, I'm trans, transman, FtM" because that's part of my past that I don't want to pretend doesn't exist.

Statements from four participants in the Early Transition Group illustrate how members of this group approach self-labeling:

> Basically [I identify] as man or male, but if I'm in the right context I'll say FtM or transman ... I prefer FtM to female-to-male, just because I hate to have that word female in there, but that's just personal preference.

> The one term I will never use is transgender ... My gender has been rock solid male from day one. It has not wandered anywhere, it has not trans'd anywhere ... The only thing that has trans'd is my anatomical sex ... From surgery on, I stopped identifying myself as transsexual ... I'm a man.

> I would say that my gender is male. I would say my gender identity is probably mostly male. I would say my anatomical sex is mostly male and the part of me that's FtM is kind of like the part of me that's Mexican—it's just part of me.

> I'm a transsexual because I've had surgeries.

Sexual Orientation

Sexual orientation varied greatly between groups. Six Recent Transition Group participants (in contrast with only two of the Early Transition Group members) reported being attracted to both men and women (and, for some, being attracted to transgender persons).

Quotations from the interviews illustrate how bisexuality is spoken about in these interviews from the Recent Transition Group:

> I identify as pansexual. So, I'm interested in anyone who plays with gender. I really like gender. I'm [attracted to] men, women, transgender.

> I identify as queer. I'm physically attracted to all different types of genders. I get into romantic relationships with male identified people.

and from the Early Transition Group:

> I guess I would say [I am attracted to] women and occasionally men.

> I identify as bi, so [I am attracted to] pretty much everybody.

Five of the Early Transition Group participants reported being primarily attracted to women, while only one member of the Recent Transition Group was primarily attracted to women. In addition to being attracted to women, one participant in each group was attracted to feminine characteristics. One member of each group was primarily attracted to men.

DISCUSSION

Over time, the experience of being "man enough" seems to have changed for FtMs. Compared to FtMs who transitioned more than 22 years ago, FtMs transitioning today (and recently) less often seek bottom surgery, use a greater variety of terms to label their gender identity, and more often identify themselves as attracted to people in more than one gender category. These cohort differences suggest that, rather than being purely anatomical/biological, the circumstances in which an FtM experiences himself as "man enough" are also shaped by the sociocultural period in which he transitions.

Body Transitions

The two cohorts were markedly similar in their near unanimous choice of hormone therapy and top surgery. However, they differed in their willingness to surgically remove internal reproductive organs or alter external genitalia.

Hormone Therapy

All FtMs of both cohorts chose hormone therapy. This suggests that the masculinization process resulting from hormone therapy feels like a critical transition step in order for FtMs to feel "man enough." The ensuing changes (e.g., lowering of the voice, increased body and facial hair, and increased muscle mass) provide an FtM with his first opportunity to observe his gender identity being reflected in his new male physical characteristics. In addition to his experience of himself, these physical changes create experiences through which he is identified by others as male/man and is treated as male/man.

Top Surgery

Similarly, all participants included (or desired) top surgery as a part of their transitions. This suggests that, for FtMs to feel "man enough," they want to perceive their chests as flat and without breasts. In addition to this essential perception of one's own body, results of chest surgery may also allow an FtM additional ways to enjoy his body and to be perceived as male by others in a larger number of settings, for example, baring his chest at the beach and changing his shirt in the locker room. The burden of binding, covering, and hiding his chest is alleviated by top surgery.

Bottom Surgery

Genital reconstructive surgery can provide an FtM with the final components of his male body: male genitalia (Ettner, 1999). Traditionally, this change was expected to bring greater satisfaction with his male body and also with his body in relationship to another during intimate and sexual encounters. This transformation can also provide relief for an FtM in other contexts, for example, using a public urinal.

All members of the Early Transition Group had some form of bottom surgery, and the majority had genital reconstructive surgery. Perhaps all members of this cohort felt that, despite the arduous, complex recovery and the less than optimal outcomes, they wanted to take this step in order to feel "man enough." At the same time, by taking this step, they were transitioning according to the path set forth by the early gender clinics, which included bottom surgery. During this era, if an FtM expressed ambivalence about any of the steps to transition (including bottom surgery), he may have been deemed ineligible and terminated from the program (Cromwell, 1999; Meyerowitz, 2002). This may have created conditions in which FtMs did not feel able to consider each choice separately and in relation to their own individual preferences and needs.

Bottom surgery is less important to FtMs transitioning today; in fact, no participants from the Recently Transitioned Group made this choice. FtMs in the 21st century seem to be saying, "Yes, I am man enough" without any type of bottom surgery.

There may be some practical considerations at work that affect this choice. Many FtMs cannot afford the high costs associated with these surgeries, and insurances rarely offer assistance (Cromwell, 1999; Rubin, 2003). In addition, the results of such surgeries are inconsistent and frequently unsatisfying. Therefore, even if an FtM can afford the operations, he is not sure he wants to buy something he may (or may not) receive.

However, the pervasive decision to defer or choose against bottom surgery appears to be based on a finding that some find startling and confusing: FtMs are satisfied with their bodies. Earlier, the prevailing view held that bottom surgery was imperative for successful transition and that any FtM who does not desire bottom surgery should not be eligible for transition (Cromwell, 1999; Meyerowitz, 2002). That the Recent Transition Group participants are satisfied with their gender identity, expression, and body shows that bottom surgery is not a necessary conclusion of transition for all FtMs and that a person's satisfaction with his body does not mean he is not a "true transsexual" or "transsexual enough." Some FtMs answer yes to the question, "Am I man enough yet?" without any type of bottom surgery.

That symptoms of gender dysphoria appear to dissipate with the physical changes that result from hormone therapy and chest surgery provides evidence for the centrality of gender presentation, or "doing gender" (Butler, 1990) in redressing these feelings. Being able to be perceived "as a man" by most people, under most circumstances, may be felt as the essential component of being "man enough"; a phallus may not be at the heart of this concern.

Self-Labeling

In the early days, professionals (and many transgender people, too) assumed a binary view of gender: male or female (Meyerowitz, 2002). Following from this binary view, FtMs were presumed to want to disidentify from the sex assigned at birth (female) and to label themselves with the words used to identify the other or "opposite" sex/gender (male, man). Findings from this study reveal that it has never been quite so simple. Most participants of both groups use a variety of terms to express their gender identity. Using different labels allows an FtM to maintain a multifaceted gender identity that includes all the components with which the person identifies.

Advocates of the binary perspective would predict that FtMs would prefer to self-label as male. Indeed, all members of the Early Transition Group used this label and the majority of the Recent Transition group did so. Some FtM participants did not want to identify with anything that indicated they had been female previously. At the same time, a majority of participants in both cohorts sometimes labeled themselves with terms that acknowledged a sense of identity as "trans." By doing this, they indicate that they do not always want to be perceived as "simply male," but sometimes want to claim a gender identity that acknowledges their history and transition. In the recent cohort, there is greater evidence of a "trans" identity as an end point, not just as a transition point in passage from one side of the binary to the other.

All self-labels included some identification with maleness. Even FtMs who described themselves as transgender, transsexual, or genderqueer imply maleness because that is what one has transitioned toward. For FtMs, to be "man enough" is to incorporate some maleness into their identity and labels for self. FtMs' attempts to answer "Am I man enough yet?" in the self-labeling process involve choosing (or creating) terms to differentiate and identify one's gender, and making deliberate choices about the circumstances in which to use a particular term.

Those who had transitioned since 2000, used a greater variety of gender identity terms. This may be related to the availability of more terms at the time of transition. More resources are currently available for FtMs transitioning—the Internet, literature, organizations, support groups, and gender therapists all can provide an array of differentiated identity terms (Ettner, 1999; Lev, 2004).

Sexual Orientation

Sexual orientation, like gender identity, was characterized by expanded possibilities in identity and interests among those who transitioned in the 21st century. The majority of the Recent Transition Group was attracted to both men and women (and, for some, transgender), while the majority of the Early Transition Group was primarily attracted to women.

Members of the Early Transition group, when navigating the journey to physical transition, were likely to have experienced a great deal of scrutiny about their gendered goals. In order to be approved for transition, they had to persuade medical and mental health "gatekeepers" of their "manly" intentions. For some professional practitioners (and perhaps for some FtMs, too) a man who was sexually attracted to other men may not have seemed "manly" enough at a time in our cultural history when stereotypes of gay men confounded a gay male sexual orientation and femininity, and gender was in part defined by sexual orientation. Today, our cultural understandings have become less simplistic and stereotypic (Cromwell, 1999). In addition, FtMs preparing for transition have access to others who have gone before and, through the Internet as well as through personal and organizational contacts, are able to get advice about what these gatekeepers "need to hear" (Lev, 2004). FtMs in the 21st century may have more room to distinguish between what they "need to say" to get through the gate, as distinct from their own subjective desires. Moreover, the awareness of diversity in sexual orientation and gender identity has dramatically increased among the health professionals who serve the role of gatekeeper, and a bisexual or same-gender sexual orientation is no longer considered a contra-indication for sex reassignment (Bockting, Benner, & Coleman, 2009).

Unlike earlier cohorts of FtMs, those who transition today (or recently) are often connected to GLBT communities (Cromwell, 1999). Perhaps because the Recent Transition Group has had more access to gay-, lesbian-, bisexual-, and queer-identified people, they have explored their sexual orientation more fully. This exploration, validated by the GLBT community, may allow a sense of fluidity about sexual orientation and also a recognition of attraction to gender traits and roles (masculinity, femininity) independent of sex (the male-female continuum). With increased awareness and education about sex, gender, and sexuality, people are now more able to articulate their attractions to others and have a community that supports and validates this.

A part of answering "Am I man enough yet?" is considering one's self in relation to potential sexual partners and sexual acts. For some, being man enough may be related to fully realizing one's sexual attractions and romantic and relational interests. Although gender identity is distinct from sexual orientation, FtMs may experience their sexuality more fully once they experience a sense of congruence with their gender presentation. Feeling one's self to be a "man among men" or a man in a (sexual/romantic) relationship to a woman may be a key component of satisfaction (Sammons, 2010).

As contemporary FtMs are showing through their physical changes, gender identity terms, and sexual orientation, the route of transition should not be prescribed ahead of time, without knowing the effects that each change will have on an individual's identity and sexual development.

Limitations

Findings from this study are informed by a small sample of 16 people. These (mainly Caucasian and educated) participants may differ from FtMs of other ethnic or class backgrounds. In addition, most participants were recruited from transgender and FtM community spaces, and were involved in the transgender community and "out" about their transition; FtMs that live primarily by stealth were less likely to participate.

The retrospective self-report method that was chosen is likely to render results that reflect participants' felt truth about their experiences; however, some participants' memories may be limited or distorted.

The study may also have been influenced by the first author's (Adam F. Yerke's) disclosure to the participants of his own FtM transition.

Hopefully, this commonality between researcher and participant provided a sense of safety to participants, allowing them to be more open and use jargon. However, some participants may have modified their remarks based on their beliefs about the researcher and his experience.

Conclusion

This study underscores the impact of historical-cultural time for an FtM's choices for physical transition, self-labeling terms, and expressed sexual interests and attractions—shaping his identity and lifestyle as a man, transgender person, and individual. FtMs, and the clinicians and psychotherapists working with them across generations, may need to address the implications, for that individual, of having transitioned during a time when few transgender people were represented in the media or when hate crimes against transgender people were peaking. As psychotherapists become aware of the ways that historical context shapes their clients' lives, they are able to have a deeper understanding of the clients and so are in a better position to help clients make decisions that are authentic, recognizing that this authenticity has been influenced by the sociocultural time period in which clients transitioned. After all, an FtM is ultimately transitioning to be a more authentic being, whether he decides that he is "man enough" as a man, as a transman, or with another new identity.

NOTE

1. The term *man* is used here to represent the aim of transition for FtMs. Although not all FtMs identify as men, this term is used here to represent the satisfaction that one hopes to gain by transitioning, no matter his gender labels.

REFERENCES

American Psychiatric Association. (2000). *Diagnostic and statistical manual of mental disorders* (4th ed., Text rev.). Arlington, VA: Author.

Benjamin, H. (1966). *The transsexual phenomenon*. New York, NY: Julian Press.

Bockting, W. (2008). Psychotherapy and the real-life experience: From gender dichotomy to gender diversity. *Sexologies, 17*, 211–224.

Bockting, W. O., Benner, A., & Coleman, E. (2009). Gay and bisexual identity development among female-to-male transsexuals in North America: Emergence of a transgender sexuality. *Archives of Sexual Behavior, 38*, 688–701.

Bolin, A. (1988). *In search of Eve: Transsexual rites of passage*. South Hadley, MA: Bergin & Garvey Publishers.

Brown, M. L., & Rounsley, C. A. (1996). *True selves: Understanding transsexualism—For families, friends, coworkers, and helping professionals*. San Francisco, CA: Jossey-Bass.

Butler, J. (1990). *Gender trouble*. New York, NY: Routledge.

Califia, P. (1997). *Sex changes: The politics of transgenderism*. San Francisco, CA: Cleis Press.

Coleman, E., Bockting, W. O., & Gooren, L. J. G. (1993). Homosexual and bisexual identity in sex-reassigned female-to-male transsexuals. *Archives of Sexual Behavior, 22*, 37–50.

Cromwell, J. (1999). *Transmen and FtMs: Identities, bodies, genders, and sexualities*. Chicago: University of Illinois Press.

Devor, A. H. (1997). *FTM: Female-to-male transsexuals in society*. Indianapolis: Indiana University Press.

Etaugh, C. A., & Bridges, J. S. (2004). *The psychology of women: A lifespan perspective*. Upper Saddle River, NJ: Allyn & Bacon.

Ettner, R. (1999). *Gender loving care: A guide to counseling gender-variant clients*. New York, NY: Norton.

Feinberg, L. (1999). *Transliberation: Beyond pink or blue*. Boston, MA: Beacon Press.

Gooren, L. J. G. (1999). Hormonal sex reassignment. *International Journal of Transgenderism, 3*(3). Retrieved from http://www.symposion.com/ijt/ijt990301.htm

Green, J. (2000). Look! No, don't!: The visibility dilemma for transsexual men. In S. Stryker & S. Whittle (Eds.), *The transgender studies reader* (pp. 499–509). New York, NY: Routledge.

Israel, G. E., & Tarver, D. E. (1997). *Transgender care: Recommended guidelines, practical information, and personal accounts*. Philadelphia, PA: Temple University Press.

Kotula, D. (2002). *The phallus palace*. Los Angeles, CA: Alyson.

Lev, A. I. (2004). *Transgender emergence: Therapeutic guidelines for working with gender-variant people and their families*. Binghamton, NY: Haworth Press.

Lothstein, L. M. (1983). *Female-to-male transsexualism: Historical, clinical and theoretical issues*. Boston, MA: Routledge.

Meyer, W., Bockting, W. O., Cohen-Kettenis, P., Coleman, E., DiCeglie, D., Devor, H., ... Wheeler, C. C. (2001). The Harry Benjamin International Gender Dysphoria Association's standards of care for gender identity disorders—Sixth version. *International Journal of Transgenderism, 5*(1). Retrieved from http://www.symposion.com/ljt/soc_2001/index.htm

Meyerowitz, J. (2002). *How sex changed: A history of transsexuality in the United States.* London, UK: Harvard University Press.

Miles, M. B., & Huberman, A. M. (1994). *Qualitative data analysis.* Thousand Oaks, CA: Sage.

Pauly, I. B., & Edgerton, M. T. (1986). The gender identity movement: A growing surgical-psychiatric liaison. *Archives of Sexual Behavior, 15,* 315–329.

Rachlin, K. (1999). Factors which influence individual's decisions when considering female-to-male genital reconstructive surgery. *International Journal of Transgenderism, 3*(3). Retrieved from http://www.symposion.com/ijt/ijt990302.htm

Rachlin, K., Green, J., & Lombardi, E. (2008). Utilization of healthcare among female-to-male transgender individuals in the United States. *Journal of Homosexuality, 54,* 243–258.

Rubin, H. (2003). *Self-made men: Identity and embodiment among transsexual men.* Nashville, TN: Vanderbilt University Press.

Sammons, D. (2010). *Body beautiful: The impact of body image on sexual pleasure in a transgender population* (Unpublished doctoral dissertation). Alliant International University, San Francisco, CA.

Socarides, C. W. (1970). A psychoanalytic study of the desire for sexual transformation [transsexualism]: The plaster-of-paris man. *International Journal of Psycho-Analysis, 51,* 341–349.

Stryker, S., & Whittle, S. (Eds.) (2006). *The transgender studies reader.* New York, NY: Routledge.

Tsoi, W. F., Kok, L. P., Yeo, K. L., & Ratnam, S. S. (1995). Follow-up study of female transsexuals. *Annals of the Academy of Medicine, Singapore, 24,* 664–667.

Valentine, D. (2006). "I went to bed with my own kind once": The erasure of desire in the name of identity: In S. Stryker & S. Whittle (Eds.), *The transgender studies reader* (pp. 407–419). New York, NY: Routledge.

Weiss, R. S. (1994). *Learning from strangers: The art and method of qualitative interview studies.* New York, NY: The Free Press.

Yerke, A. F. (2008). *The transition experiences of two cohorts of female-to-male transsexuals: The impact of socio-cultural context* (Unpublished doctoral dissertation). Alliant International University, San Francisco, CA.

Gender Expression as a Reflection of Identity Reformation in Couple Partners Following Disclosure of Male-to-Female Transsexualism

Christine Aramburu Alegría
Deborah Ballard-Reisch

ABSTRACT. This qualitative study examined gender expression as a reflection of identity reformation in male-to-female transsexual persons and their nontranssexual female partners following the disclosure of transsexualism in relationships that were established as man-woman. Each partner in 17 couples participated in individual interviews. The gender expression of each couple partner evolved as male-to-female persons progressed in their male-to-female transition. Three themes related to male-to-female gender expression emerged as the couples reformed their identities and relationships: (a) phases: exploration/experimentation, adolescence, and maturation; (b) (re)gendering as women: behavior and interactions; and (c) visual congruence: dual/single presentation and sex reassignment surgery.

KEYWORDS. Gender identity, transsexual, gender expression, couples, qualitative research

In recent years, transsexual persons have collectively increased their visibility in society. The movies *Normal* (Brokaw & Pilcher, 2003) and *Transamerica* (Macy & Tucker, 2005) and the media coverage of a female-to-male (FtM) transsexual person giving birth to children (Childs, 2011) are recent examples that bring attention to diverse identities. Studies and personal reports on transsexualism can also be found. These studies and personal reports often examine transsexual lives from a biomedical perspective (e.g., Cohen-Kettenis & Gooren, 1999; Moore, Wisniewski, & Dobs, 2003; Santos-Ocampo, 2007), a focus on lived experiences of individual transition (e.g., Mason-Schrock, 1996; Morgan & Stevens, 2008), the challenges and adjustments partners and families of transpersons experience (Alexander, 2003; N. R. Brown, 2009; Hines, 2006), and strategies to help in the counseling of these individuals and families (Israel, 2004; Lev, 2004; Raj, 2008). Taking a social-psychological perspective, this study contributes to the body of knowledge on transsexualism through its examination of the relational lives of transsexual persons and their primary partners. We use data from a larger study (Aramburu Alegría, 2008) on relational activities and identity reformation to examine gender expression in male-to-female (MtF) transsexual persons and their nontranssexual natal female (i.e., female since birth; F) partners in relationships that had been established as

---

Christine Aramburu Alegría is affiliated with the Orvis School of Nursing at the University of Nevada Reno in Reno, Nevada. Deborah Ballard-Reisch is affiliated with the Elliott School of Communication at Wichita State University in Wichita, Kansas.

Address correspondence to Christine Aramburu Alegría, University of Nevada Reno, Orvis School of Nursing, MS 0134, Reno, NV 89557, USA. E-mail: aramburu@unr.edu

man–woman. For ease of reading, the term "female" or the abbreviation "F" will be used to denote the nontranssexual natal female partner.

Using in-depth, individual interviews with each partner from 17 MtF–F couples, we examine gender expression to illuminate how identities and relationships are reformed to maintain (a) individual coherence, signifying congruence between behavior, physicality, and identity, and (b) relational coherence, signifying congruence between partner interactions and relational views (Baldwin, 1992; Fiske & Taylor, 1991; Holmes, 2000). In this manner, a contextual vantage point from which to discuss the various ways MtF transsexual persons and their female partners experience their relationships as MtF persons and assume more authentic lives is offered. Although empirical data are lacking, the general belief is that couple relationships dissolve following the disclosure of a partner's transsexualism. The present study examines gender expression and identity reformation in couples who have made a decision to continue and to invest in the reformation of their relationships.

TRANSSEXUAL IDENTITY AND GENDER EXPRESSION

Traditional notions of sex and gender conflate the two and create expectations that gender expression be based on the assigned gender category, which is based on assigned sex. This categorization typically occurs at birth, when the child's sex is assigned as male or female, based on external genitalia (Butler, 1990). From that point on, according to Butler (1990), the construction and expression of gender is a life-long process, typically involving seeing, learning, and behaving in expected ways. For many MtF persons, this expectation has led to longstanding incongruence between identity and demonstrable gender signifiers (M. L. Brown & Rounsley, 1996; Gagné & Tewksbury, 1998; Morgan & Stevens, 2008; Vitale, 2001).

More contemporary conceptions of sex and gender disequate the two, allowing for the recognition of gender variance (e.g., Bornstein, 1994). Gender identity and expression can be fluid, and its expression can comprise a range of characteristics and behaviors from masculine to feminine, both, or neither. This fluidity can manifest from day to day, and individuals may elect not to self-assign gender and, indeed, may refute the actual notion of gender (Andre & Guitierrez, 2010; Kusalik, 2010; Luengsuraswat, 2010). Moreover, genitalia need not be conceptualized as a signifier of gender. That is, a woman or man may have a penis, vagina, or intersex genitalia. For example, in a study by Bockting, Benner, and Coleman (2009), some FtM transsexual persons were able to successfully find coherence between their natal genitalia (i.e., vagina) and their identity, while others desired phalloplasty. Importantly, for some transsexual persons, transition and expression need not be a linear progression from one gender to the other. Rather, they may oscillate between genders, identifying feminine and masculine aspects within themselves and demonstrating these characteristics outwardly (Diamond & Butterworth, 2008).

Various models of transsexual identity development have been described and include initial phases of self-examination and possible subsequent disclosure to others (Bockting & Coleman, 2007; Lev, 2004). During these early phases of identity development, the transsexual person may or may not be able to identify or use precise language for feelings of incongruence (Bockting & Coleman, 2007). Further, from an early age, transsexual persons engage in personal decision making on matters related to authentic gender expression and disclosure. Issues of stigma, rejection, and discrimination force many to live lives of inauthentic gender expression (Bockting & Coleman, 2007; Vitale, 2001). Transsexual persons who do disclose their identity will then often embark upon exploration and experimentation toward finding personal comfort and coherence within their newly emerging identity (Bockting & Coleman, 2007; Lev, 2004). Steps are taken to increase signifiers of femininity or masculinity to create authentic self-expression (Gagné, Tewksbury, & McGaughey, 1997; Morgan & Stevens, 2008) and to present with increasing conviction (Dozier, 2005; Schrock, Reid, & Boyd, 2005). In turn, enacting a convincing gender expression and receiving affirming feedback from others signifies achievement and further validates and authenticates transsexual persons' gender identity (Crawley, 2008; Dozier, 2005; Iantaffi & Bockting, 2011; Rubin, 2003). During this phase of exploration and experimentation, gender expression may be highly exaggerated and stereotyped. As transsexual identity development progresses, gender expression becomes more appropriate and a less salient aspect of the transsexual person's life. During this phase of identity integration, greater confidence and congruity between self-view and public persona has been achieved by the transsexual person (Bockting & Coleman, 2007; Lev, 2004).

MtF persons who aim to increase feminine signifiers often learn behaviors associated with womanly embodiment through the observation of symbolic others. Examples include viewing and interacting with other MtF persons in the media, on the Internet, or in chat rooms (Gagné et al., 1997). Social gatherings that involve transsexual persons present other opportunities through which feminine expression may be learned. Through these measures—interacting with present or symbolic others—an MtF person observes the gender expression of a woman, and through these representations construct the representation of her own womanhood. Through this dynamic, the self-view as a woman emerges and evolves. Thus, for partnered MtF persons who do not disclose their gender identity within relationships that were established as man–woman, two processes exist. The first is the intrapersonal MtF identity, shaped outside of the couple relationship; the second is the interpersonal identity, acting according to the relational norms the couple has established as a man–woman relationship. Both representations—MtF, an internal representation as woman, and man, an external representation—are shaped via interactions, direct or indirect, with present or symbolic others (Blumer, 1969; Schrock et al., 2005; Swann, 2005).

Following the disclosure of transsexualism within a relationship that was established as man–woman, these norms and representations stand to change as partners interact to regain relational cohesion (Blumer, 1969; Holmes, 2000). According to Lev's (2004) Family Emergence Stages, negotiation often follows a period of turmoil in which confusion is paramount and relational norms disrupted. Following negotiation, successful adaptation and recreation of relational norms is possible (Lev, 2004).

TRANSSEXUAL IDENTITY, GENDER EXPRESSION, AND RELATIONSHIPS

Heteronormativity often guides the relational norms and gender expression of partners in couples that include transsexual persons. For example, female partners of FtM persons have reported looking forward to their partners' "top surgery" (i.e., removal of breasts) so that their partners could achieve congruence between masculine gender identity and presentation. Moreover, they positively anticipated their partners' top surgery so that their own views of their partners as men could be facilitated (Pfeffer, 2008). Similarly, MtFs' alternating presentations of man and woman have been found to be unsettling and highly stressful for their female partners (Erhardt, 2007). In this not uncommon phase of MtF transition, the transsexual person's partner may positively anticipate the completion of transition. These findings suggest that some partners of transsexual persons may have a greater preference for what may be viewed traditionally as physical congruence—and a lesser preference for gender fluidity—within their partners. Further, these findings reflect efforts to negotiate and navigate the relationship toward a renewed equilibrium (Lev, 2004).

Within relationships, gender expression extends beyond the transsexual person's physicality; it impacts each partner's identity and behaviors and the couple's interactions (Holmes, 2000). For example, to advance their partners' sense of masculinity, female partners of FtMs have also reported interacting with their partners in more culturally defined feminine ways (Kins, Hoebeke, Heylens, Rubens, & De Cuypere, 2008; Pfeffer, 2008; Ward, 2010). Further illustrating the effect of one partner's physical transformation on the identity of the other partner, N. R. Brown (2009) found that lesbian-identified partners of FtM persons reported a decrease in the salience of their own identities as sexual-minority women as their partners increased their masculine presentation. Voice and language are also notable gender signifiers. In a study of 20 couples in which each partner is transsexual, Ettner (2007) found that the language used by FtM participants was less emotive than that of MtF participants, a pattern that is often more common in natal males than in natal females.

Day-to-day routines in nontraditional couples can also diverge from those of traditional couples. In a study examining FtM–F relationships, Pfeffer (2010) found that female respondents, despite viewing themselves as feminists, performed the majority of the household labor, a phenomenon demonstrated by many traditional heterosexual couples (see Lachance-Grzela & Bouchard, 2010, for review). On the other hand, following MtF transition, heteronormative behavior has been discarded by MtF–F couples who report greater equality in running their households (Erhardt, 2007; Hines, 2006).

In addition to modifications in interpersonal dynamics that occur with changing relational contexts, the body dysphoria that accompanies FtM identity has been shown to have a negative effect on the self-view of females in FtM–F couples (Pfeffer, 2008). Female partners of FtM persons who are in the process of FtM transition have described increasing self-examination and self-consciousness regarding their own bodies, suggesting that females' self-view is influenced by their partners' self-views.

In sum, these studies demonstrate the contextual nature of gender expression within couple relationships and its influence upon, and effects from, partner interactions as they engage in the continual dynamic of relationship negotiation. The present study expands upon the existing body of work on transsexual gender identity and nontraditional relationships by examining gender expression as a reflection of identity within the context of MtF–F relationships that were established as traditional man–woman relationships. Specifically, we examine two questions:

1. How do MtF and F expressions of gender evolve in MtF–F relationships following disclosure of transsexualism and during MtF transition as partners renegotiate their relationship?
2. How does this evolution of gender expression affect partners' identity reformation, and conversely, how does identity reformation affect gender expression?

METHOD

To uncover rich detail that can lead to greater understanding, this study uses a qualitative methodology. Rather than producing findings that will be generalizable to all MtF–F couples, the study's goal is to provide insight into the gender expression of MtF–F couple partners as they reform their identities following the disclosure of transsexualism and subsequent gender transition. It is important to note that these couples were established as traditional man–woman relationships and that they have remained intact following disclosure of transsexualism.

Participants

Seventeen MtF–F couples (34 individuals) from five states (Arizona, California, New Mexico, New York, and Washington) were recruited via Internet forums and network sampling. The Internet forums were moderated and focused on the education and support of transgender persons and their significant others, friends, families, and allies (SOFFAs).

All MtFs self-identified as "female," "transsexual woman," or "transwoman." Participation in the study was voluntary and without reward or payment. Inclusion criteria included (a) participation of both partners, (b) F partner assignation as female sex at birth, (c) MtF assignation as male at birth and self-identification as transsexual, (d) partners' commitment to each other, and (e) relationship's duration of at least 1 year. With the exception of one cohabitating and engaged couple, all couples were married and living together. The participants ranged in age from 34 to 67 ($M = 53.6$ years). Approximately three fourths of the participants were between the ages of 50 and 67 (F: $n = 12$, 70%; MtF: $n = 13$, 76.4%). More than one half of the participants ($n = 23$, 67.6%) had a bachelor's degree or higher level of education. All participants described themselves as White/Caucasian.

The length of the couples' relationships ranged from 3 to 44 years ($M = 21.8$ years). At the time of the disclosure of transsexualism, the couples had been in their relationships from 1.5 to 35 years ($M = 16.2$ years). The average length of time since the disclosure ranged from 5 months to 12 years ($M = 5.6$ years). The stage of transition among the MtF participants ranged from presenting as woman part-time ($n = 3$, 17.6%) to presenting as woman full-time ($n = 14$, 82.4%). The extent of MtFs' body-altering procedures ranged from hair removal only to complete transition with sex reassignment surgery. Among those who were living full-time as women, almost one half had had sex reassignment surgery ($n = 6$, 42.8%), and approximately one fourth ($n = 3$) had SRS scheduled. All MtF respondents reported the use of medications for feminization. Table 1 lists participant age, length of relationship at time of disclosure, amount of time since disclosure, and MtF feminization at the time of data collection.

Procedures

Data Collection

Intact MtF–F couples were invited to participate in a study focusing on relationship activities and identity reformation following disclosure of MtF transsexualism. Data collection continued until thematic saturation was achieved (Creswell, 1998; Lincoln & Guba, 1985; Miles & Huberman, 1994). Individual semistructured interviews were conducted with each participant (see Table 2 for interview questions). With the exception of one couple, all interviews were face to face. Due to logistics involving distance and travel, the interviews with each member of this couple were conducted individually by telephone. An interview protocol consisting of 15 open-ended questions was developed. These questions elicited information on the couples' relationships and identity prior to and following the disclosure of transsexualism. With the exception of the phone interviews with one couple, the first author traveled to all participants' homes, and conducted, audiorecorded, and transcribed verbatim all interviews. Measures were taken to ensure each respondent's privacy during the interview. The duration of each interview was approximately 1.5 to 2 hr. Thirty-four interviews were completed, resulting in approximately 59.5 hr of interview and 522 pages of transcription.

Data Analysis

Inductive methods were used to identify themes related to gender expression in identity reformation. The first author and a second coder independently used the constant comparative method, an iterative process by which data are compared within and across subjects (Glaser & Strauss, 1967; Strauss & Corbin, 1998). Consistent with this approach, the analysis of the data was focused with the following questions: "What is going on here in terms of their relationship and their identity?" and "How does gender expression relate to their relationship and identity?" With these questions in mind, the researchers first read through each transcript in its entirety to get an overall sense of the content of the interview. Then, line-by-line coding was used to identify themes. Dominant themes and subthemes were grouped according to conceptual linkages.

To increase confidence in the study findings, intercoder consensus and verification strategies were employed (Creswell, 1998; Lincoln & Guba, 1985; Miles & Huberman, 1994). The researchers independently coded data from three randomly chosen couples for themes related to identity reformation. On completion of the independent codings, the researchers met and reviewed the codings for the six interviews (three MtF and three F). Adjustments were made to the themes via consensus and a single list of themes was created for use in coding the remaining interviews. The six interviews were corrected with codes based on the revised consensus typology.

The researchers independently coded the data from the remaining 14 couples. As new themes emerged, they were negotiated via e-mail exchange and face-to-face meetings to ensure consistent, comparable use of thematic categories. After all data were coded, each coder independently reviewed the coded data to ensure that any codes added/changed during the process were consistently applied to previously coded

TABLE 1. Participant age, Length of Relationship, Length of Relationship at Time of Disclosure, Length of Time Since Disclosure, and MtF Feminization ($N = 34$)

Subject	Age	Length of relationship	Length of relationship at time of disclosure	Time since disclosure	Feminization
F1	54	37 years	27 years	10 years	
MtF1	54				Full-time, SRS
F2	62	44 years	35 years	9 years	
MtF2	66				Full-time, SRS
F3	54	17 years	6 years	11 years	
MtF3	54				Full-time, SRS
F4	63	13 years	10 years	3 years	
MtF4	56				Full-time, SRS
F5	50	16 years	12 years	4 years	
MtF5	55				Full-time, scheduled for SRS
F6	57	22 years	13 years	9 years	
MtF6	65				Part-time, hair removal
F7	53	3 years	1.5 years	1.5 years	
MtF7	50				Full-time, hair removal
F8	46	20 years	15 years	5 years	
MtF8	48				Full-time, scheduled for SRS
F9	34	14 years	13.5 years	5 months	
MtF9	38				Part-time, hair removal
F10	45	13 years	10 years	3 years	
MtF10	48				Full-time, scheduled for SRS
F11	67	25 years	13 years	12 years	
MtF11	63				Part-time, hair removal
F12	60	22 years	20.5 years	1.5 years	
MtF12	67				Full-time, hair removal, orchiectomy
F13	58	23 years	19 years	4 years	
MtF13	55				Full-time
F14	49	30 years	27.5 years	2.5 years	
MtF14	50				Full-time, hair removal, tracheal cartilage shave, vocal cord surgery, discussing SRS
F15	61	39 years	30 years	9 years	
MtF15	63				Full-time, SRS
F16	38	6 years	1 year	5 years	
MtF16	38				Full-time, hair removal
F17	50	27 years	22 years	5 years	
MtF17	51				Full-time, SRS

Note. F = female; MtF = male-to-female; SRS = sex reassignment surgery.

data. Following this review, the researchers met and reviewed and discussed the themes until consensus was reached.

Verification

The "trustworthiness" (Lincoln & Guba, 1985) of the study was increased through methods outlined by Lincoln and Guba (1985) and Creswell (1998). These include (a) member checks and (b) community member checks. These methods enrich the credibility of the study and increase the opportunity for previously overlooked themes to be newly expressed. Member checks with participants allowed them to validate the researcher's interpretation of

TABLE 2. Interview Questions

1. What first attracted you to your partner? (NF & MtF)
2. Since disclosure, has the relationship changed? How? (NF & MtF)
3. Do you communicate/interact differently with your partner since the disclosure? (NF & MtF)
4. Has your partner's transsexualism led you to make any adaptations within your relationship? (NF) Has your transsexualism led you to make any adaptations within your relationship? (MtF)
5. Are the behaviors or activities that you use to maintain your relationship now different from the behaviors or activities that you used prior to the disclosure of transsexualism? (NF & MtF)
6. What types of interactions or activities within the relationship have you stopped since the disclosure? (NF & MtF)
7. What types of interactions or activities within the relationship have you started since the disclosure? (NF & MtF)
8. What types of interactions or activities within the relationship have been modified since the disclosure? (NF & MtF)
9. Does transsexualism add any unique challenges/opportunities to your relationship? (NF & MtF)
10. Has the way you see yourself changed since the disclosure? (NF & MtF)
11. Has the way you see your partner changed since the disclosure? (NF & MtF)
12. Has the way you see the relationship changed since the disclosure? (NF & MtF)
13. Some people would say that you are now in a same-sex relationship. What does that mean to you? (NF & MtF)
14. How would you describe your relationship with your partner before the disclosure of transsexualism? (NF & MtF)
15. How would you describe your relationship with your partner after the disclosure of transsexualism? (NF & MtF)

Note. NF = natal female; MtF = male-to-female.

their experiences. This was conducted during all interviews with participants, and with 10 participants (five couples) following their interviews. During the interviews, respondents' information was frequently validated by repeating back to them what they seemed to be reporting. For example, the statement "What I am hearing you say is ____. Is that correct or is it off?" was used during the interview process. This allowed respondents to rectify and/or clarify the researcher's interpretation. Participants were also asked to elaborate on their responses. At the end of each interview, participants were asked, "Is there any issue or topic missing or being overlooked? Is there anything you want to add?" The couples were also allowed the opportunity to communicate with the interviewer at any point following the interview. Five couples (i.e., 10 individuals [30%]) participated in post-transcription member checks. All verified that the study findings described their experiences.

Findings from the study were also reviewed with two nonparticipant members or allies of the transsexual community. These individuals included an MtF individual who is a peer counselor of other MtF individuals and a nontranssexual counselor who works with the transsexual community. Both verified the findings as fitting to the community.

RESULTS

Three themes of gender expression reflecting partners' identity reformation emerged: (a) phases: exploration/experimentation, adolescence, and maturation (100%, $n = 34$); (b) (re)gendering as women: behavior and interactions (100%, $n = 34$); and (c) visual congruence: dual/single expression and sex reassignment surgery (F: 76.5%, $n = 13$; MtF: 82.4%, $n = 14$). These themes express the essence of the evolution of gender expression in MtF–F relationships (RQ1) and how this evolution affects identity reformation (RQ2). The participants described MtFs' evolution as women as one that progressed from disclosure, through an exploration of and experimentation with gender, to finally finding their authentic presentation. Along this often nonlinear trajectory, the gender expression of the partners evolved and was modified as partners' interactions and identities similarly evolved.

Phases: Exploration/Experimentation, Adolescence, Maturation

The process of identity reformation and (re)learning gender as women began with a period of exploration and experimentation. This exploration and experimentation included seeking resources on the Internet; frequenting gay, lesbian, and genderqueer establishments; and experimenting with alternative sexual lifestyles in order to seek a fit between gender expression and gender identity.

Phases of MtF respondents' gender expression included (a) exploration and experimentation, (b) adolescence, and (c) maturity. The period of exploration and experimentation was a period of searching for the true or authentic nature of the MtFs' gender. Couple partners described a period of experimentation, exploration, and self-examination during this process. Exploration was described as a journey by the respondents, including MtF17, who compared her quest to seek coherence between her appearance and her identity to a train ride. She described taking select stops along the way to experiment with different ways of presenting, until finally she found where she needed to be as her authentic self:

> I liken it to a train ride, going on a journey. You're on this train, and you're going for a trip, and you get off a stop, and you look around and you think "You know what, this just isn't for me," and you go a little further, and then "You know what, I think I need a little more." It's like when I first tried on women's underwear. It was ok, but it didn't make me feel whole. Then you go a little further, try putting on a little makeup and it's still not there. I ended up getting off at the last stop, including the sex reassignment surgery. (MtF17)

Often this initial period of exploration and experimentation with gender expression included the belief that the MtFs' identity may only involve an interest in women's clothing (i.e., cross-dressing). An initial belief that their partners were "simply" cross-dressers allowed female partners an introduction to cross-gender activity and a period of time to adjust to changes in the relationship. F17, whose partner compared the transition to a train ride, described her initial belief that her partner's gender was limited to an interest in women's clothing. She was able to accept cross-dressing as long as there were limits on it:

> I thought mainly it was a fascination with female clothing. I never realized how deeply embedded it was although I knew she wasn't happy being the way she was, being a man. She would get together with other cross-dressers, and for a while I thought—and she did too—that that was enough. I was fine with it. It was confined and limited, and that's what I needed—to believe it was about the clothes and that it was confined. (F17)

Respondents also reported a period of sexual experimentation in an effort to explore different ways of expressing sexuality and gender, in hopes that the experimentation would satisfy the MtFs' need for an alternate gender expression. Experimentation included engaging in alternative sexual practices and social networks, such as venues where gender norms could be challenged. For example, F8 described a year of sexual experimentation in order to explore her partner's gender identity:

> It was actually a dominant–submissive group—female dominant–male submissive—because we were thinking that maybe he needed to be more submissive at home, and maybe it was just a matter of looking at the female role in society. Maybe that was what he needed when he wasn't in the high stress job at work. His job and his work are always so high stressed. Maybe that was where we needed to go with our relationship and then this would all go away. It was like, "Well maybe if we do this, this will be what's right and it won't necessarily have to be about being a woman, maybe it's just a role," and so we tried that club and we got to know people who considered themselves cross-dressers and transsexuals. We met dominatrix in the community, and we went to classes about dominant and submissive roles and how to fill those roles. (F8)

All of the couples ultimately found that the authentic identity of the MtF partner was transsexual. Following the couples' period of exploration and experimentation, the process of MtFs' un-gendering themselves as men and re-gendering themselves as women began. This process included a series of phases in which MtFs sought to present themselves in a manner that would reflect their self-view as women. Finally enjoying the liberation of their true selves, MtFs utilized cultural standards of beauty as models by which to gender themselves. Hence, their initial presentation as women often reflected the idealized self-view they had long privately held of themselves as young, beautiful, and sexy women, regardless of their chronological age. Respondents termed this phase as "adolescence," or, as MtF16 put it, "the Britney Spears" phase. This phase was not well received by female partners. MtF13 illustrated:

> It was all so new for me. Like being a teenager—a teenage girl. I had to grow up and get through that because it was driving [F13] crazy. I would want to go out dressed inappropriately, or not really dressed for my age, or wear things that are too tight. I would be really giddy and not take things seriously. (MtF13)

During this and subsequent phases, MtFs relied on their partners' advice on how to perform as women. This included advice on dress, mannerisms, makeup, and safety. Assisted by their partners, The MtFs' confidence and presentation as women eventually became more age and situation appropriate, and no longer bound by unrealistic media images. With time, they viewed themselves as having "grown into" themselves as mature women. A typical example is provided by MtF13, who went on to describe her ability to finally mature from her teenage phase into the woman she is:

> I was trying all these different things on. I had to mature. We went from being husband and wife, to being like a mother and daughter, to being two mature women. I finally took ownership. I had to get over myself. I took ownership of being a transwoman. I settled into myself. (MtF13)

Observing and advising their partners in their evolving gender expression caused many female partners to examine their own identity as women and found this process to be a period of growth. Female partners freely acknowledged that their MtF partners were the *"femmes"* in the relationship, and that they were indeed now the more "masculine" ones. Yet, despite perceiving an increase in their own masculinity, none of the female partners in the present study reported altering their gender presentation in order to highlight their own femininity. Rather, they enjoyed yielding the feminine spotlight to their partners, whom they viewed as having waited their entire lives to be their true selves. Indeed, female partners reported finding liberation and self-affirmation, and a release of the need to embody what they believed society expected of them as women. F12 stated:

> For a period of time I felt kind of butch because, really, [MtF12] is more feminine than I am. I don't wear makeup and jewelry and all that stuff, and now she's taken to primping and getting earrings. She's very interested in feminizing herself. I completely understand, and I think it's fine, but I'm the opposite now. I don't need to shave my legs, but she does. So I went through this period of feeling like maybe I am a butch person. And that's ok. She is the feminine one. It's her time. But it helped me examine myself and really accept myself. (F12)

Similarly, F16 reported that her partner's gender transition caused her to examine dimensions of her own identity and gender expression that she previously had suppressed:

> There is hard stuff in this for me, in terms of my own identity and sexuality, and really it was stuff I would have rather not dealt with, but I knew I had to. Society doesn't make it easy. I had to deal with my own sense of my own masculinity, in terms of role. I now feel less scared of playing the masculine role. I like being able to play more the guy's role in a relationship, and I've never really dealt with that before. (F16)

(Re)gendering as Women: Behavior and Interactions

Recognizing the increasing confidence and validation in their MtF partners as women, female partners sought to further educate them through advice on behaviors expected of women. For example, MtF10 described her self-view of her feminine mannerisms and the assistance her partner gives her:

> [F10] has shown me a lot. I like it when people recognize [my feminine ways]. For instance, the way I hold my hands, the way I look at my fingernails. Female. When I drop something, guys just bend at waist and pick it up. I stoop down. There were about 30 [feminine] mannerisms I learned. These were things that made me feel comfortable. Now, the thing [F10] gets mad about is if I don't practice my girl voice. (MtF10)

As the couples' relationship as two women evolved, partners dispelled previously assumed expectations and customary activities. They no longer viewed a need to perform relational activities according to what they had perceived as typically expected of men or women. Prior to disclosure, MtFs were resistant to performing activities they did not consider to be "men's" activities, such as household chores. These were modified as a result of changed expectations. These modifications were welcome as they made relationships more equitable and cooperative. MtF17 provided this example:

> I hated, never mind being the guy, having to portray the macho, decision-making leader that I thought a guy had to portray. Absolutely this relationship has now changed totally. It's run by both of us. That's the way it should have been from day one. I was trying so hard to play the guy role, so anything that even slightly would resemble a feminine thing, I resisted big time. Wash the floor? Forget it! Now, the house gets done. I have no problem—cooking, laundry—it's a totally different relationship. We're equals and make decisions together. That has been a big change in the relationship. (MtF17)

Interactions indicative of affection were also modified. For example, F16 described modifications in displays of affection:

> [MtF16] had this thing when she was out, she would put her hand on my shoulder when she was presenting as a guy. It's kind of a masculine thing to do, and she doesn't do that anymore with me. I do it more with her, because our gender roles have switched in that sense. I tend to be the more protective, possessive one now, certainly the one playing the more butch role, such as who opens doors and stuff. Acting that way makes more sense to us now. (F16)

As MtFs gained validation and confidence as women, they were able to resume some of the masculine activities that they had previously enjoyed as men. For example, MtF13 described returning to auto repair as her hobby, stating, "When I started transition, I quit working on cars. I didn't think women did that stuff. Now, I know I'm an attractive woman. Working on cars doesn't change that."

Visual Congruence: Duality/Singleness, Sex Reassignment Surgery

As transition began, to help maintain relational stability and identity coherence, a duality of MtFs' visual presentation was prominent for the couple partners. Specifically, the process of MtFs' gender expression within the transition led to the following experiences: (a) dual gender expression, to be followed by single expression and (b) seeking congruence between the genitals and gender identity.

Initially, MtFs limited their feminine presentation to certain days, times, or situations, presenting as men at other times. Female partners' readiness to interact with their MtF partners as women was often the determining factor in the extent of their partners' feminine gender expression. This alternating expression allowed the female partners to gradually adjust to their partners' re-gendering as women, until such time as they were ready to view their partners as women full time. In the following example, F1 provided a typical description of the necessity of viewing her partner as a woman initially for only short periods of time:

> A big thing for me was seeing her around the house in female clothes. There were certain times of day that she did it. Just short periods at first. That was a big thing that took time but it helped me a lot to kind of actually experience whether I could deal with it or not, so it really played in the "Could I really live with the person of the female gender? Could that really be ok for me?" So, at first, she dressed as a woman for short periods, then those periods grew. (F1)

In the process of gender duality to gender singleness, a period of "waiting and catching up" was often necessary, especially for female partners. Slowing the pace of transition was necessary to catch up emotionally and cognitively with the situation. Partners required time for their relational views to "settle" before further progression in transition could occur. F2 illustrated:

> I'd need time to catch up with the rapid changes we were going through. It involved communication and reassurance. For example, [MtF2] might say, "I'd like to pierce my ears. Are you ready for that?" And I might say, "No, let's wait on that." Or she might say, "I'd like to cross-dress today, are you ok with that?" Just her asking, "Are you ok?" and accepting that if I say "No," it isn't that I am never going to say yes and it isn't that I hate her. It's just that I'm tired. I'm mentally exhausted. That's a huge issue. That's part of catching up. Slow down. Let me catch up my mental image of where we're at, so I can deal with it. (F2)

Yet, as MtFs' alternating man/woman gender expression continued, the adjustment required of the female partners as they interacted with their partners' presentations created confusion and frustration. What was initially adaptive for identity coherence and reformation—limiting feminine expression—later became maladaptive. As F6 stated, "[i]t became too confusing, to see a man leave for work in the morning and a woman coming home at night." Hence, a solely feminine presentation became possible as the female partners increasingly viewed their MtF partners as women.

Female partners described experiencing a sense of loss as they realized the life they envisioned with their MtF partners would not come to pass. Their partners' alternating presentations as men and as women made repeatedly salient the loss of their partners as men. Consequently, the majority of the couples made the decision to abandon gender expression duality, and to instead move forward with a consistent feminine presentation. This example reflects F1's view of her partner as a woman, as well as her readiness to give up her partner as a man:

> I actually got to the point where it was like, "I can't deal with this anymore." I told her, "You are one person at home and you're another person that goes off to work and I've just had it. Do one or the other. I'm done. I cannot deal with this *two-person*." It was upsetting and confusing. I think that part of it was that I needed to know what my life was going to be like, and that was the point at which it really became ok with me for her to go out in the world [as a woman]. We weren't going to be the husband and wife who grew old, or whatever I thought it was supposed to be like. It wasn't going to be that way, and to be confronted visually became harder and harder for me. It was like, "If I'm going to let go of it, I need to not see it. I'm done, I don't want to be reminded five days a week every morning that I've lost that." I'm willing to lose it but not have it thrown in my face everyday. That's where I got to in my thinking. I needed to live my life on one side of this or the other. I couldn't keep going back and forth. (F1)

Duality in presentation also increased dissonance and frustration for MtFs. In this example,

MtF1 describes her increasing identification as a woman as time spent in feminine expression increased:

> I went from being a guy at work to being androgynous at work. I would be androgynous at work but female at home. At first that felt great. But over time the androgynous state got harder. It's very hard to get up in morning and put on clothes that could be either/or [in gender] when it's not what you want to be, not really how you want to dress. I started having problems with that, and the longer I straddled that line the harder it got to not be female all the time. I was feeling more and more female as time went on. (MtF1)

The positive personality changes that female partners observed in their MtF partners when presenting as women contributed to the couples' decision to abandon MtFs' expression as men. Female partners reported that their MtF partners were more positive and pleasant to interact with when presenting as women, and increasingly angry and withdrawn when presenting as men. Likewise, MtFs felt more content and validated as women as their time in feminine expression increased. Thus, the couples' joint decision to go forward with full-time feminine presentation became clear for both partners. F2 provided this example:

> It was much easier to live with someone that's a woman. It was hard going back and forth between these two people because [MtF2] was so different from [who she is as a man]. [As a man,] she was uptight and angry and controlling, not fun, depressed. [MtF2] was cheerful and happy. I was also going back and forth between relating to a man and to a woman. I was relating to this uptight ugly type of person and this really joyful person. The joyful person was the one I wanted to live with but it was very scary and very hard, and the uptight one was the one that looked like the person I had been married to but actually didn't seem a lot like the person I had married. It was just so hard for her to continue acting like a man. (F2)

Three MtFs reported presenting as women less than full time, and that issues related to disclosure prevented their readiness to present full time as women. MtF6 reported that she had not disclosed to her elderly mother next door, MtF9 to her young children, and MtF11 to fellow attendees at the symphony.

The majority of the couples' gender expression reformation also included efforts to establish what they perceived as culturally normative congruity between the genitals and gender identity (i.e., women have vaginas). Nine (52.9%) of the 17 MtFs reported having had or being scheduled for sex reassignment surgery, and one MtF and her partner were in the process of actively planning it. Despite full-time gender expression as women—in the home, at work, and in public—MtF respondents reported that having sex reassignment surgery was still the ultimate goal, or as MtF8 described it, "the holy grail." However, due to various reasons (e.g., health and/or financial), not all MtFs could attain this goal. In this example, MtF8 described the disconcerting aspect of looking at her reflection in the mirror and her need for sex reassignment surgery:

> The deep-seated reason [for sex reassignment surgery] is that it's the "holy grail." It's something I need to do in order to be able to look back in the mirror, coming out of the shower. Right now I kind of have selected vision gaps. I see from the waist up and from the knees down and I don't think about what's in between. So in order to be comfortable that way, I need that part of my anatomy to not be the way it is. It needs to be the other design. It's going to make me feel more whole. (MtF8)

Female partners also reported the need to have what they perceived as congruence between their partners' genitals and gender identity. Seeing their partners with what they perceived to be female genitalia enhanced their ability to view their partners as women and helped reduce the dissonance associated with physical incongruence (i.e., woman with penis). In the following example, F4 described her and her partner's decision making regarding sex reassignment surgery and how the surgery facilitated her evolving view of her partner as a woman:

> [MtF4] did not like the penis even being there, and for me it was too confusing as well, mentally and emotionally, to see this person as a woman and then seeing her penis. So we decided to let go of that. It was too confusing for both of us. She had the surgery, and that helped me see her as a woman. It helped a lot. Seeing her as a woman was hard for me. Even though she was working very hard at it, I could see the man in the dress. The dressing was of a woman, but my mind had not yet projected or really believed that she was really a woman. So, the surgery helped with that. (F4)

DISCUSSION

As in previous research, the findings of the present study demonstrate the contextual influence on gender expression (Kins et al., 2008; Pfeffer, 2008; Ward, 2010). We first briefly discuss the findings in terms of the transsexual individual, and then discuss the findings within the context of relationships.

Transsexual Identity and Gender Expression

In opposition to their self-view, the MtF participants in the present study presented themselves as men in interpersonal situations prior to disclosure. Following disclosure of transsexualism to their partners, the MtFs' process of gender expression and identity reformation began with gender experimentation and exploration, findings that are consistent with previous studies and current models of transsexual identity formation (Bockting & Coleman, 2007; Hines, 2006; Lev, 2004; Wester, McDonough, White, Vogel, & Taylor, 2010). Consistent with previous research by Schrock and colleagues (2005) that found an association between "womanly" embodiment and validation as women, MtFs in the present study felt increasingly validated and authentic as women as they increased their time in feminine expression. In many instances, feminine gender expression was exaggerated and stereotypical, such as MtF16's description of her "Britney Spears phase." Only when their gender expression was convincingly feminine could they relax the demonstration of feminine signifiers and resume typically masculine activities that they had previously enjoyed as men. These findings are consistent with previous research and also support the concept of identity integration as a later stage of identity development (Bockting & Coleman, 2007; Dozier, 2005; Lev, 2004).

In contrast to findings that suggest that the identity and gender expression of transsexual persons may oscillate between two genders (Diamond & Butterworth, 2008; Roen, 2002), MtFs in the present study sought to continue a direct trajectory toward womanhood. Similar to FtMs and other MtFs (Pfeffer, 2008; Vitale, 2001), the MtFs in the present study reported displeasure with viewing their bodies. Indeed, the MtFs reported striving for complete transition, including sex reassignment surgery, to achieve complete congruency between physicality and identity. In this regard, the MtFs in the present study diverge from transsexual persons who demonstrate greater gender identity fluidity, such as those described in Diamond and Butterworth's (2008) study in which four self-identified FtM persons reported that they do not need to remove physical signifiers of femaleness in order to take on masculine characteristics and identity.

Transsexual Identity, Gender Expression, and Relationships

In the present study, participants demonstrated the evolution of their gender expression toward achieving congruency between identity, gender expression, and relational dynamics. In this phase of negotiation (Lev, 2004), couples in the present study worked at navigating their relationship toward renewed stability. Couples communicated on matters of transition, such as its pace, and worked out mutually acceptable solutions. F2 illustrated this when she stated to her partner, "Slow down. Let my mental image catch up." In this manner, couples were able to make their (often nonlinear) way toward the next stage, finding balance, as described by Lev (2004).

When the present study findings are compared to previous studies of couples that include a transsexual person, much convergence can be found. Just as female partners of FtM persons looked forward to their partners having "top" surgery (Pfeffer, 2008), female partners of MtFs in the present study were ultimately supportive of their partners' sex reassignment surgery. Further, similar to previous findings (Erhardt, 2007), as their comfort with MtF expression as women increased, female partners experienced increasing dissonance with their partners' duality as man/woman. This dissonance extended into physicality and female partners expressed perceiving visual incongruity in their partners' physical being prior to sex reassignment surgery. Just as feminine embodiment increases self-validation as women for MtFs (Schrock et al., 2005), female partners' view of their partners as women also increased following sex reassignment surgery. Further, female partners' statements such as "If she is going to be a woman, then I want her to be a nice-looking, classy woman" (F2) and "It was hard, trying to see her as a woman when she had a penis" (F3) illustrate their need to have their partners conform to cultural heteronormative expectations of women in dress and in body. As in previous research (Ettner, 2007), these expected and desired feminine gender signifiers also included an attractive feminine voice. Indeed, to use a popular phrase, for MtFs and their partners who educate them on feminine matters, the devil is in the details.

Important to recall is that the couple partners established their relationships as man–woman. The MtFs were careful to present as men in their relational interactions prior to disclosure, and relationship norms were negotiated as such. Findings may be different for MtF–F couples who established their relationships with the realization of each other's authentic identities. It is possible that couples who enter nontraditional relationships have different views on what constitutes congruency of identity, physicality, and gender expression. For example, Factor and Rothblum (2008) found that persons of sexual and gender diversity were attractive to genderqueer persons.

Consideration should also be given to generational and other socially constructed ideologies that may have shaped the study participants' views regarding physical embodiment of gender. The mean age of the participants was more than 50, and almost three quarters of the participants were between 50 and 67 years of age. Comfort with sex and gender fluidity may be limited in persons who, generationally speaking, have had less exposure to sexual and gender diversity than younger generations, who may more readily embrace diversity. It should be noted, however, that day-to-day household routines were modified following the MtFs' transition. In this regard, similar to previous research (Erhardt, 2007; Hines, 2006), couples in the present study reported that household chores, once generally the domain of gender stereotypical roles, became more egalitarian.

More specific to their gender expression, female partners of MtFs welcomed an increase in masculinity to increase their partners' femininity and validation as women. These findings are similar to female partners of FtMs who modify their gender expression to increase their partners' masculinity and validity as men (Kins et al., 2008; Pfeffer, 2008). However, unlike females in FtM–F couples who intentionally accentuate their femininity to enhance their partners' gender validity, female partners in the present study did not report deliberately masculinizing themselves. Rather, they reported greater self-acceptance as women who no longer need to adhere to cultural norms regarding femininity. In this regard, FtM–F couples may be viewed as moving toward heteronormative norms, while MtF–F couples may increasingly diverge from heteronormativity.

The greater self-acceptance reported by the female partners in the present study is in contrast to the decreased self-acceptance reported by female partners in FtM–F couples, who reported increasing self-consciousness with their bodies (Pfeffer, 2008). Collectively, these findings suggest that female partners' self-view and gender expression are influenced by their transitioning MtF and FtM partners. MtFs' and FtMs' self-view and gender expression become salient in this situation and in turn influence their partners' own self-view. For female partners of FtMs, the female body may become a source of self-consciousness, yet feminine embodiment may be increased to validate their partners' masculinity (Pfeffer, 2008). For female partners of MtFs, the female body may become more accepted, and feminine embodiment may not need to be emphasized.

PRACTICAL IMPLICATIONS

The findings from this study offer insights into MtF–F couple relationships via in-depth interviews with 17 MtF–F couples. The findings cannot be generalized to all MtF–F couples, and clinicians need to be cognizant of and open to the many diverse presentations of gender. Nonetheless, the study findings can offer clinicians and others who work with the transsexual community another look into the gender expression and identity reformation that can occur in couple relationships that stay intact following disclosure of transsexualism. Relationship maintenance is a process, and while the factors that determine whether or not a couple stays together can be myriad, the present study offers a look into an aspect of couple relationships that have thus far been successful in reforming their relationships. The successful consensual negotiation of gender expression may be one factor that enhanced these couples' odds of staying intact. The experiences of these study participants can provide clinicians a perspective from which they may compare and contrast their personal observations. The experience of their own clients may or may not be similar.

LIMITATIONS AND FUTURE DIRECTIONS

Study limitations can provide opportunities to gain further understanding in future studies. As such, they are identified. First, the sample size is small. Although this allowed for in-depth individual interviews, it limits generalizability of the findings. Second, the participants in this study comprised a relatively homogenous sample. All identified as White/Caucasian, more than half had completed a bachelor's degree or higher level of education, and the majority were over 50 years of age. To gain a more comprehensive understanding of the relationships of transsexual persons, future research should sample a variety of ethnic, racial, and socioeconomic populations across age groups and geographic and cultural boundaries. Third, most of the couples in the present study had been together a relatively long time prior to the disclosure of transsexualism, and further, for most of the couples, more than 3 years had passed since the disclosure. Study findings could be different for couples who had not been together as long at the time of disclosure, for couples who had not had as much time lapse since the disclosure, or for couples who did not wish or plan to stay together.

Fourth, recruitment was conducted through Internet forums and network sampling. Random sampling of the population is not possible, and these nonrandom methods of recruitment represent a selection bias, further limiting generalizability of the findings. Internet forums required access to computers and the Internet, and network sampling allowed individuals who are "out" and acquainted with one another to be referred to the study. The descriptions of the participants' experiences may not be representative of couples who do not participate in Internet forums or have access to other contacts within the community. Finally, future studies utilizing quantitative measures may further enrich understanding of gender expression and identity in transsexual persons and their relationships.

FINAL COMMENT

Without exception, participants reported that they hoped a greater understanding of transsexualism and relationships would result from the study findings. In that spirit, this study offers another insight into the various ways MtF transsexual persons and their female partners experience their relationships. The encouraging findings demonstrate that gender expression can be modified and heteronormative norms challenged in the successful reformation of identity and relationships. Via the increasing visibility of transsexual persons, and a greater number of studies examining issues affecting this population, it is our hope that the social marginalization and discrimination that force so many into lives of suppression will decrease.

REFERENCES

Alexander, J. (2003). There are different points in your life where you can go either way. *Journal of Bisexuality, 3*(4), 129–150.

Andre, A. P., & Guitierrez, L. (2010). In our skin. In K. Bornstein & S. B. Bergman (Eds.), *Gender outlaws: The next generation* (pp. 157–162). Berkeley, CA: Seal Press.

Aramburu Alegría, C. (2008). *Relational maintenance and schema renegotiation following disclosure of transsexualism: An examination of sustaining male-to-female transsexual and natal female couples* (Doctoral dissertation). Available from ProQuest Dissertations and Theses database. (UMI No. 3316374).

Baldwin, M. W. (1992). Relational schemas and the processing of social information. *Psychological Bulletin, 112*, 461–484.

Blumer, H. (1969). *Symbolic interactionism: Perspective and method*. Englewood Cliffs, NJ: Prentice-Hall.

Bockting, W., Benner, A., & Coleman, E. (2009). Gay and bisexual identity development among female-to-male transsexuals in North America: Emergence of a transgender sexuality. *Archives of Sexual Behavior, 38*, 688–701.

Bockting, W., & Coleman, E. (2007). Developmental stages of the transgender coming-out process: Toward an integrated identity. In R. Ettner, S. Monstrey, & A. E. Eyler (Eds.), *Principles of transgender medicine and surgery* (pp. 185–208). New York, NY: Routledge.

Bornstein, K. (1994). *Gender outlaw: On men, women, and the rest of us*. New York, NY: Routledge.

Brokaw, C., & Pilcher, L. (Producers), & Anderson, J. (Writer/Director). (2003). *Normal* (Motion picture). USA: HBO Films.

Brown, M. L., & Rounsley, C. A. (1996). *True selves: Understanding transsexualism*. San Francisco, CA: Jossey-Bass.

Brown, N. R. (2009). "I'm in transition too": Sexual identity renegotiation in sexual-minority women's relationships with transsexual men. *International Journal of Sexual Health, 21*, 61–77.

Butler, J. (1990). *Gender trouble: Feminism and the subversion of identity*. New York, NY: Routledge.

Childs, D. (2011). Pregnant man Thomas Beatie may stop at 3 kids. *Medical Unit, ABC News*. Retrieved from http://abcnews.go.com/blogs/health/2011/10/31/pregnant-man-thomas-beatie-may-stop-at-3-kids/

Cohen-Kettenis, P., & Gooren, L. (1999). Transsexualism: A review of etiology, diagnosis, and treatment. *Journal of Psychosomatic Research, 46*, 315–333.

Crawley, S. L. (2008). The clothes make the trans: Region and geography in experiences of the body. *Journal of Lesbian Studies, 12*, 365–379.

Creswell, J. W. (1998). *Qualitative inquiry and research design*. Thousand Oaks, CA: Sage.

Diamond, L. M., & Butterworth, M. (2008). Questioning gender and sexual identity: Dynamic links over time. *Sex Roles, 59*, 365–376.

Dozier, R. (2005). Beards, breasts, and bodies: Doing sex in a gendered world. *Gender & Society, 19*, 297–316.

Erhardt, V. (2007). *Head over heels: Wives who stay with cross-dressers and transsexuals*. New York, NY: Haworth Press.

Ettner, R. (2007). Transsexual couples: A qualitative examination of atypical partner preferences. *International Journal of Transgenderism, 10*, 109–116.

Factor, R. J., & Rothblum, E. (2008). Exploring gender identity and community among three groups of transgender individuals in the United States: MTFs, FTMs, and genderqueers. *Health Sociology Review, 17*, 235–253.

Fiske, S. T., & Taylor, S. E. (1991). *Social cognition*. New York, NY: McGraw-Hill.

Gagné, P., & Tewksbury, R. (1998). Conformity pressures and gender resistance among transsexual individuals. *Social Problems, 45*, 81–101.

Gagné, P., Tewksbury, R., & McGaughey, D. (1997). Coming out and crossing over: Identity formation and proclamation in a transgender community. *Gender & Society, 11*, 478–508.

Glaser, B. G., & Strauss, A. L. (1967). *The discovery of grounded theory*. Chicago, IL: Aldine.

Hines, S. (2006). Intimate transitions: Transgender practices of partnering and parenting. *Sociology, 40*, 353–371.

Holmes, J. G. (2000). Social relationships: The nature and function of relational schemas. *European Journal of Social Psychology, 30*, 447–495.

Iantaffi, A., & Bockting, W. O. (2011). Views from both sides of the bridge? Gender, sexual legitimacy and transgender people's experiences of relationships. *Culture, Health & Sexuality, 13*, 355–370.

Israel, G. E. (2004). Supporting transgender and sex reassignment issues: Couple and family dynamics. *Journal of Couple & Relationship Therapy: Innovations in Clinical and Educational Interventions, 3*, 53–63.

Kins, E., Hoebeke, P., Heylens, G., Rubens, R., & De Cuypere, G. (2008). The female-to-male transsexual and his female partner versus the traditional couple: A comparison. *Journal of Sex and Marital Therapy, 34*, 429–438.

Kusalik, T. (2010). Identity, schmidentity. In K. Bornstein & S. B. Bergman (Eds.), *Gender outlaws: The next generation* (pp. 54–60). Berkeley, CA: Seal Press.

Lachance-Grzela, M., & Bouchard, G. (2010). Why do women do the lion's share of housework? A decade of research. *Sex Roles, 63,* 767–780.

Lev, A. I. (2004). *Transgender emergence.* New York, NY: Haworth Press

Lincoln, Y. S., & Guba, E. G. (1985). *Naturalistic inquiry.* Beverly Hills, CA: Sage.

Luengsuraswat, B. (2010). Proof. In K. Bornstein & S. B. Bergman (Eds.), *Gender outlaws: The next generation* (pp. 242–246). Berkeley, CA: Seal Press.

Macy, W. (Producer), & Tucker, D. (Writer/Director). (2005). *Transamerica* (Motion picture). USA: Weinstein Company IFC Films.

Mason-Schrock, D. (1996). Transsexuals' narrative construction of the "true self." *Social Psychology Quarterly, 59,* 176–192.

Miles, M. B., & Huberman, A. M. (1994). *Qualitative data analysis.* Thousand Oaks, CA: Sage

Moore, E., Wisniewski, A., & Dobs, A. (2003). Endocrine treatment of transsexual people: A review of treatment regimens, outcomes, and adverse effects. *Journal of Clinical Endocrinology & Metabolism, 88,* 3467–3473.

Morgan, S. W., & Stevens, P. E. (2008). Transgender identity development as represented by a group of female-to-male transgendered adults. *Issues in Mental Health Nursing, 29,* 585–599.

Pfeffer, C. A. (2008). Bodies in relation—Bodies in transition: Lesbian partners of trans men and body image. *Journal of Lesbian Studies, 12,* 325–345.

Pfeffer, C. A. (2010). "Women's work?" Women partners of transgender men doing housework and emotion work. *Journal of Marriage and Family, 72,* 165–183.

Raj, R. (2008). Transforming couples and families: A trans-formative therapeutic model for working with the loved-ones of gender-divergent youth and trans-identified adults. *Journal of GLBT Family Studies, 4,* 133–163.

Roen, K. (2002). "Either/or" and "both/neither": Discursive tensions in transgender politics. *Signs: Journal of Women in Culture and Society, 27,* 501–522.

Rubin, H. (2003). *Self made men: Identity and embodiment among transsexual men.* Nashville, TN: Vanderbilt University Press.

Santos-Ocampo, A. (2007). New onset systemic lupus erythematosus in a transgender man: Possible role of feminizing sex hormones. *Journal of Clinical Rheumatology, 13,* 29–30.

Schrock, D., Reid, L., & Boyd, E. M. (2005). Transsexuals' embodiment of womanhood. *Gender & Society, 19,* 317–335.

Strauss, A., & Corbin, J. (1998). *Basics of qualitative research.* Thousand Oaks, CA: Sage.

Swann, W. B. (2005). The self and identity negotiation. *Interaction Studies, 6,* 69–83.

Vitale, A. (2001). Implications of being gender dysphoric. *Gender & Psychoanalysis, 6,* 121–141.

Ward, J. (2010). Gender labor: Transmen, femmes, and collective work of transgression, *Sexualities, 13,* 236–254.

Wester, S. R., McDonough, T. A., White, M., Vogel, D. L., & Taylor, L. (2010). Using gender role conflict theory in counseling male-to-female transgender individuals. *Journal of Counseling and Development, 88,* 214–219.

RANDAL F. SCHNOOR AND MORTON WEINFELD

Seeking a Mate:
Inter-Group Partnerships among Gay Jewish Men

ABSTRACT/RESUME

While documentation and discussion of out-marriage rates among heterosexual Jews is widespread, very little similar analysis exists for inter-group partnering patterns of gay Jews. Using both quantitative and qualitative data, this study addresses this void. The 2001 Canada Census data reveal that Canadian gay and lesbian Jews are "out-marrying" at a rate of approximately eighty-nine percent. This can be compared to an out-marriage rate of approximately thirty percent for Canadian heterosexual Jews. A qualitative sociological study on gay Jewish men both corroborates the census finding that gay Jews out-marry at a very high rate and offers a rich and nuanced account of the forces behind this statistic. We found that the majority of the gay Jewish men interviewed expressed a desire for a Jewish partner, but only a very small proportion of these men actually had Jewish partners. Several reasons are cited to explain this discrepancy. Findings about gay Jews are compared to both heterosexual Jews and to other minority gay men.

Tandis que la documentation et la discussion des taux de mariage exogroupe parmi les juifs hétérosexuels sont répandues, il y a un manque d'analyse semblable pour des modèles de partenariat entre les groupes pour les juifs gais. En utilisant des données quantitatives et qualitatives, cette étude vise à combler ce manque. Les données du recensement du Canada de deux mille un indiquent que les juifs gais et lesbiens canadiens pratiquent le mariage exogroupe à un taux d'approximativement quatre-vingt neuf pourcent. Ceci peut être comparé à un taux de mariage exogroupe d'approximativement trente pour cent pour les juifs hétérosexuels canadiens. Une étude sociologique qualitative sur les hommes juifs gais corrobore le recensement constatant que les juifs gais pratiquent le mariage exogroupe à un taux très élevé, et offre un compte-rendu riche et subtil des forces qui se trouvent sous cette statistique. Nous avons constaté que la majorité des hommes juifs gais qui ont été interviewés a exprimé un désir pour un partenaire juif, mais seulement une proportion très petite de ces hommes a eu réellement des partenaires juifs. Plusieurs raisons sont citées afin d'expliquer cet écart. Des résultats au sujet des juifs gais sont comparés aux juifs hétérosexuels et à d'autres hommes gais minoritaires.

INTRODUCTION

This study examines partnering practices among a small sample of gay Jewish males, a double minority. The specific topic of concern is the extent to which these partnerships involve Jews and non-Jews and the factors associated with exogamous or endogamous relationships. Questions of Jewish identity in particular will be explored as these relate to the partnership decisions of gay Jews. Comparisons will be made with heterosexual Jews as well as gay individuals of other ethno-religious minorities.

Out-Marriage in the Jewish Community

While the issue of the extent to which gay Jewish men form inter-group partnerships has never been systematically examined, much attention has been devoted to the question of out-marriage rates among Jewish heterosexuals (out-marriage is defined as a Jew who marries a non-Jew who does not convert to Judaism). Researchers in the United States have demonstrated that the rates of Jewish out-marriage have climbed continuously from marriages in the 1940s (of which approximately 7 percent were out-marriages), to marriages in the 1960s (approximately 17 percent), to marriages in the 1970s (approximately 30 percent) (Cohen 1988, 28–30; Sharot 1998, 95–96). The statistic that most startled the American Jewish community was the finding of the 1990 National Jewish Population Study (NJPS). The survey reported that, from the period between 1985 and 1990, some 40 to 50 percent (depending on one's interpretation of the statistics) of American Jews were married to non-Jews (Fishman 2000, 141).[1] Recent figures for Canada tell us that in the mid-1990s, Canadian Jews were out-marrying at a rate of approximately 30 percent (Weinfeld 2001, 154). It is important to contextualize these out-marriage rates. In Canada, the Jewish rates are lower than most other religious groups. For example, Catholics out-marry at a rate of 42 percent; East Orthodox at 50 percent (Weinfeld 2001, 372).

Some scholars argue that Jewish out-marriage may be a positive development for the Jewish community. For example, Jewish out-marriages can be seen as encouraging non-Jews to embrace Jewish culture (Weinfeld 2001, 154), to be creating, in total, more "Jewish households" (households with at least one Jew) (Cohen 1991, 49), and as a signal of a diminishing of animosity or mistrust between Jews and non-Jews (Fishman 2002, 9–11). Many leaders of the Jewish polity argue, however, that out-marriage is one of the most serious problems facing the Jewish community of North America today because, overall, it threatens the primary goal of "Jewish continuity." Jewish out-marriage is reported to be particularly dangerous to the Jewish community because it brings with it a loss of Jewish identity, a weakening of social and religious cohesiveness of the community, and an actual demographic loss over time of the Jewish population (Goldstein 1992, 124). A central reason offered why Jews are out-marrying is that they place less emphasis on parochial community concerns and more emphasis on the autonomy of the individual to make his or her own romantic choice based on love (Fishman 2000, 141; Weinfeld 2001, 160).

Inter-Group Partnerships among Minority Gay Men and Lesbians

In the social scientific literature that deals with the gay experience, little discussion presently exists on the subject of the ethno-religious composition of gay partnerships. Bell and Weinberg (1978, 85) reported that the majority of their gay Black sample claimed that more than half of their romantic partners had been White. Two decades later, Peplau, et al. (1997, 21) found that "committed interracial relationships occur at a higher rate among African-American lesbians and gay men than among Black heterosexuals." Greene (1995, 103-04) reported that lesbians of colour have a greater tendency to partner with women who are not members of their ethno-religious group than do White lesbians. In addition, Jackson and Sullivan (1999, 4-5), along with Nakajima, et al. (1996, 573), report the tendency in Australia for gay Asian men to partner at high rates with Anglo-Australian gay men.

Some possible reasons given by the above researchers for these higher rates of inter-group partnerships among minority lesbians and gay men include limited pools of partners to choose from (particularly when some gay men or lesbians are "in the closet") and a lack of social infrastructure for these gay men and lesbians to meet others within their ethno-religious community, particularly outside of large urban areas (Greene 1995, 103-04; Peplau, et al. 1997, 20-21). In the case of gay Asian men in Australia, the reasons concern issues of social stratification based on ethnicity, which will be explained in more detail later in this work.

Same-Sex Common-Law Relationships

The 2001 Canadian Census provides data for the first time on Canadians who report themselves in common-law, same-sex relationships. The census did not ask about sexual orientation. Therefore, the data on same-sex partnerships should not be interpreted as a precise count of the number of gays and lesbians in Canada. Some may be living alone or in other types of family arrangements. Others may not have been comfortable in revealing their same-sex relationship through the Census. Despite these limitations, the 2001 Census provides useful quantitative data for the purposes of this study. A total of 34,200 same-sex couples were counted in Canada in 2001. This represents 0.5% of all couples. There were slightly more male same-sex couples (55%) than female (45%) (Statistics Canada 2002, 4). On an individual basis, of the 21, 449, 805 adults (21 years old and over) in Canada, 67,405 report being in a common-law, same-sex relationship (0.31%). Jews in Canada show a similar proportion. Of the 267,430 adult Jews[2] in Canada, 1,035 report being in a common-law, same-sex relationship (0.39%) (Statistics Canada 2004a).

We can also learn from 2001 Census data the rate at which Canadian Jews are involved in common-law, same-sex relationships with non-Jews. This can be compared to the rate at which Canadian heterosexual Jews out-marry, reported above to be approximately 30 percent. Of the 660 reference persons[3] on the 2001 Census who are Jewish and in a same-sex, common-law relationship, only 70 partnered with a fellow Jew. Thus, 590 out of 660 gay and lesbian Jews did not partner with a fellow Jew. This corresponds to an "out-marriage" rate of 89 percent [4] (Statistics Canada 2004b).[5]

METHODOLOGY

The qualitative data collected for this study corroborates the census finding that gay Jews out-marry at a very high rate, in addition to offering a rich and nuanced account of the forces that lie beneath this statistic. The qualitative component of this study encompasses in-depth interviews with thirty gay Jewish men in Toronto, as well as interviews with six parents of gay Jewish men. Interviews were conducted between September 2000 and August 2001.

Interview respondents were acquired through snowball sampling, through introductions at gay Jewish events, and through referrals from Jewish community networks (friends, family, acquaintances). Gay Jewish respondents were distributed evenly by age, with roughly equal numbers in their teens, twenties, thirties, forties, and fifties. One respondent was in his eighties. Respondents were also well-distributed by religious background. One third of the respondents (ten) could be described as traditional Jews, while two-thirds (twenty) could be described as liberal Jews.[6] The six parents interviewed ranged in age from fifty to sixty-nine. Three were traditional Jews and three were liberal Jews.

Certain biases exist in the sample. In order to learn first-hand about the history and developments of the gay Jewish community of Toronto, gay Jewish men who have been actively involved, both past and present, in this community were over-represented. These respondents likely place more emphasis on their Jewish identities than the average gay Jew. In addition, it is important to contextualize the Toronto Jewish community. Social scientific research of the Jews of North America has demonstrated that Canadian Jews, in general, are more traditional in their Jewish behaviours than American Jews (Shaffir and Weinfeld 1981, 13; Brodbar-Nemzer et al. 1993, 43), and Toronto (and Montreal) Jews are more traditional than Jews of other Canadian cities (Davids 1998, 212–14).[7]

Interviews were semi-structured. The same groups of central themes were explored with all respondents. Respondents were first asked to describe their Jewish and gay identities. This was followed by specific questions on whether or not respondents have a serious partner, whether they have a preference for a Jewish or non-Jewish partner, and why this may be the case. Parent respondents were asked to describe their own Jewish identities and discuss whether or not they have a preference for their gay son to partner with a fellow Jew.[8]

With the exception of four interviews where notes were taken, interviews were recorded on audiotape and later transcribed verbatim. Respondents signed an informed consent form before the start of the interview and were assured that the information gathered would be kept confidential (pseudonyms would be used in the study).

As heterosexual Jewish men, the researchers of this study can be characterized as "partial insiders" (Sherif 2001, 438) to the experience of gay Jewish men. Before interviews began, these aspects of the interviewer's identity were reported to respondents in order to foster an open and respectful environment where respondents felt comfortable sharing aspects of their personal lives.

FINDINGS

The qualitative component of the study asked two specific questions on ethno-religious partnering patterns: what preferences do the gay Jewish respondents express in terms of the ethno-religious identity of a potential partner, and what are the actual ethno-religious identities of the partners of respondents?[9] Let us turn now to the first of these questions.

Desire for Jewish Partner

Of the thirty gay Jewish men interviewed, two-thirds of the respondents reported that if they were to be involved in a long-term relationship, they would prefer their partner to be Jewish. Weinfeld (2001, 144) provides some evidence, based on personal ads in

Canadian dailies, that most heterosexual Jews in Canada also prefer to partner with fellow Jews.[10] It is not surprising to see that of the ten traditional gay Jews interviewed in this study, nine of them are among this group of men who express a desire for a Jewish partner. A traditional Jew in his thirties, Alan, admitted that "[i]f I find someone who is gay and Jewish, the interest goes up a notch, for sure. If they were a seven, they become an eight... It's a factor that would draw them to me closer. If you are gay and Jewish, you can get at least two dates!"

The most common reasons offered by respondents for their desire for Jewish partners had to do with issues of cultural affinity. A large proportion of these respondents spoke about their desire to find a partner who shares their "Jewish values," emphasizing the importance of the family, ethnic heritage, and ethno-religious symbols such as Jewish food, celebration of Jewish holidays, and the like. There is no reason to believe that these factors are any different than the factors considered important by heterosexual Jews. A secular Jew in his thirties, Shaun, described his discomfort with a non-Jewish partner who did not share his "Jewish sensibilities:" "One of the issues that I had with my last partner was, because he never really had a lot of exposure to the Jewish community, he didn't understand the importance of the holidays... and didn't understand that, let's say, for Passover and the High Holidays, that it was important for him to come back with me [to my parents' house]. He didn't understand what that impact was, whereas I knew how important it was for me to be up at his parents' on Christmas Eve. And there were a lot of fights that went on over that. And I think being Jewish really provides a common ground more than anything else. [Jews] share a very similar traditional and cultural upbringing."

Traditional Jews who may be religiously observant reported added difficulties or complications with non-Jewish partners, who may not be comfortable observing Jewish Law, such as the prohibition of turning on lights or spending money on the Sabbath. In addition, some traditional Jews simply claimed that they would not even consider dating a non-Jewish man. A traditional Jew in his twenties, Saul, explained it this way: "The only reason why I would only go for somebody Jewish is because of the way I was brought up. It has nothing to do with me being gay or straight; it was always understood that I was going to private school, this is the way you do Jewish things... like I said, I'm traditional, in the sense that I would find someone who's Jewish, so we have certain things that are in common."

Another traditional Jew admitted he has only dated non-Jewish men as "more of a sexual thing, rather than a long-term partnership search thing." He expressed his opinion on the matter quite clearly: "I would never have married a non-Jewish woman had I been heterosexual. Why should I change those values as a gay Jew? I fell in love with a non-Jewish guy several years ago. And I was totally, madly in love with him. And I broke it off because of that." The issue of becoming involved with non-Jews only for sexual relationships will be elaborated upon later in this work.

Another reason identified regarding why gay Jewish men desire Jewish partners is the fear of anti-Semitism. This phenomenon occurred in isolated cases where the respondent was the child of Holocaust survivors, causing added sensitivity to prejudice against Jews. One respondent offered this explanation: "I suppose, because

I'm a child of a Holocaust survivor, I never want to be called a dirty Jew. You know, in a fit of anger, what comes out sometimes could be so hurtful that being called a dirty Jew is probably the worst thing I could ever hear. So I would never want to put myself in that position, and [my partner] being Jewish, I don't have to worry."

The majority of respondents who offered an opinion about their parents' views on the matter indicated that their parents would prefer to see their son with a Jewish partner. A traditional Jew, David, reported: "How would my parents react if I brought home a non-Jewish guy? They would have a really hard time, and I don't know how I'm going to deal with it if it comes to that. I really don't."

Arnie, who comes from a very secular Jewish family, also reported that "they'd love for me to meet someone Jewish." Four of the six Jewish parents interviewed expressed a desire for their son to find a Jewish partner. A Jewish mother of several children, including a gay teenaged son, reported that she treats all her children the same way: "I'm hoping that he will settle down, quote-unquote, and it would be with a Jewish partner. I would like that. When our eldest [heterosexual daughter] was dating, I would say, Is he Jewish? I don't see that I should change my attitudes just because they're gay. I need to let them know that it matters to me. I ask [her], and I will ask [him] when the time comes. I don't want to approach them differently. I would like to see him meet Jewish guys."

Inter-group Realities

While we have illustrated that gay Jews report a relatively strong desire to partner with other Jews, and that parents often support this, the realities of the situation do not reflect this desire. One respondent's account stands out as an important indicator on this point. A secular Jew named Lionel reported that he placed a personal ad in a Montreal newspaper stating, "Gay Jewish male seeks the same." He reported that of the numerous responses he received, not one was Jewish. As an experiment, he then placed an ad in the same newspaper reading "Gay male wrestler seeks the same." Of the many responses he received, a disproportionate number were Jewish! While we must treat this as no more than a single anecdote, we would argue that this surprising story points to the strong tendency of inter-group dating practices among gay Jews: non-Jews were the only ones who responded to a gay Jewish man, and Jews disproportionately responded to a "wrestler," who would stereotypically be perceived as non-Jewish.

The same inter-group pattern is suggested by another respondent named Jon. This respondent, who was active in Keshet Shalom, the now-defunct gay Jewish congregation in Toronto, had the impression that non-Jewish gay men considered a "Jewish husband to be the ultimate prize." Jon reported the following common scenario: "I have no idea why, but we would get people phoning up saying, 'Hi, I want to join your group. I'm not Jewish, I don't know anything about it. But I would like to meet Jewish men. Can I join your group?'" In fact, the great majority of respondents reported that they have observed that it is much more common for Jewish men to become involved with non-Jewish men, rather than with other Jewish men.

Behavioural evidence can also be provided. Of the ten respondents in the sample who had partners, eight had non-Jewish partners.[11] While one must be cautious about generalizing based on a small, non-probability sample, this finding combined with the above evidence provides some corroboration for the high rate of out-marriage among gay Jews indicated by the census data. This fact is, of course, consistent with the findings noted above, that minority gay men and lesbians tend to partner with those outside their cultural group more than do heterosexual minority individuals.

It is interesting to note that only one-third of the sample had partners at all. This can be compared to the marriage rates of heterosexual Jewish men. According to the 1990 National Jewish Population Study (NJPS) in the United States, 46 percent of Jewish men aged twenty-five to thirty-four were married (compared to 22 percent of gay men in our sample in the same age group who had a serious partner), 73 percent of Jewish men aged thirty-five to forty-four were married (compared to 50 percent in our sample), and 77 percent of Jewish men aged forty-five to fifty-four were married (compared to 33 percent in our sample) (Fishman 1993, 256).[12] The overall partner rate for gay and lesbian adults in the United States is estimated to be approximately 50 percent.[13] It appears, therefore, that the overall gay Jewish partnership rate of 33 percent found in our sample is closer to the gay partnering rate than the Jewish partnering rate. Ambert (2003, 6) argued that gay partnerships are more likely than heterosexual partnerships to dissolve due to the fact that society offers less support to those in gay relationships and values such relationships less. As will be discussed later in this work, it is possible that the Jewish community offers even less support for gay relationships than does the overall society. This may contribute to gay partnership rates involving a Jew being even lower than gay partnership rates generally. More research is needed in order to draw more definitive conclusions about these differing rates.

Analysis of Inter-group Partnering Practices

The remainder of this work explores the issue of why gay Jewish men demonstrate a strong inter-group partnering tendency. This is a significant question given the fact that two-thirds of the sample directly reported that they would prefer a relationship with a Jewish man. Where relevant, we draw parallels between the Jewish case, other minority experiences, and the heterosexual Jewish experience. Analysis of the data identified five possible reasons for this phenomenon.

a) Small Number of Openly Gay Jews

To speculate why gay Jews tend to partner with non-Jews at a high rate, the obvious factor seems to be the sheer lack of eligible, gay, Jewish men. Finding a Jewish partner can be especially difficult if one lives away from the major urban centres, where most Jews reside (Dworkin 1997, 81). Simply stated, gay men are already restricted to a small part of the male population. To then desire a Jewish partner limits one's choices quite a bit more. As several respondents reported, gay Jews suffer from being a "minority within a minority." A respondent named Kevin captures this difficult predicament: "Every time I'm in-between relationships, I say, 'Next time, I'm going to date someone who's Jewish.' And then I never actually do. There are the guys who are interested in me, and then there's the circle of guys that I'm interested in, and then there's the circle of guys that are available. That leaves such a tiny proportion of the population...that to add another criteria like, are they Jewish?...There just isn't an out, gay, Jewish community big enough for there to be a pool to select from.

Another respondent, Bruce, reported that, because of the sheer lack of numbers, even observant gay Jews are reluctantly settling for non-Jewish partners: "I know some of my most observant friends are settling down with non-Jews because, at some point, I guess, you just give in and you just say, 'Look, I don't want to be alone anymore, and I found a person whom I like very much, and they're not that Catholic or they're not that Christian...and they tolerate my schtick for Judaism,' and that's it. In that quotation, Kevin raises an important point about closeted-ness. Not only is there a relatively small pool of gay Jews to meet in any one geographical region, several respondents commented that it is even harder to find partners, because gay Jews are frequently closeted about their homosexuality. While it may be possible to find closeted gay Jews through the Internet, for some gay Jews, they are very unattractive partners. Matthew emphatically stated: "I don't ever want to date somebody who's ashamed of me. I refuse to be closeted. It's not that I'm flamboyant, but I refuse to be closeted. I don't want to live a lie. And that means that if you're not openly gay, you're not going to want to be in a relationship with me."

The negative reactions gay Jews receive from their families and other community members (Schnoor 2003, chap. 2–3) can encourage these men to keep their sexual orientation a private matter. Due to community pressure and homophobia, many Black (Peplau et al. 1997, 20) and Asian men (Chan 1989, 18–19) also keep their homosexuality hidden, thus making it difficult for other co-ethnics to find them. This leads to the important discussion of the issue of non-acceptance by the Jewish community.

b) Non-Acceptance by the Jewish Community

While small numbers of gay Jews may seem like the obvious reason why gay Jewish men do not partner with other Jews at high rates, the data reveal that lack of acceptance by Jewish families and the larger Jewish community of their gay members can have a very strong effect on whether gay Jewish men seek out other Jews. One respondent in our study, Kyle, reported that when he was in university and his parents found out he was gay, they were so upset that they cut off all financial support. What is significant is the effect this reaction had on Kyle's desire for a Jewish partner. "It was pumped into me just like most of us, marry somebody Jewish. You marry somebody Jewish; you date somebody Jewish; you should always meet somebody Jewish. When I had the kind of reaction that I had from my parents when I came out, it was clear to me that my parents were ashamed. My mother was very worried that it would reflect [badly] on her...[She] was absolutely mortified that the Jews, all those Yentas, would find out that [Max Greenblatt's] son was [gay]...So that whole experience really turned me off Jews. It really did. And I went off to the bars, and I was involved in other organizations, but nothing Jewish. I was really rebelling."

Another respondent, Jeffrey, described the challenges faced in constructing a positive gay Jewish identity and how this can decrease one's desire to find a Jewish partner. "To start with, even if you're not gay, having a Jewish identity is a challenge in the diaspora, because it's an optional thing. So you have that thing already. Then you have the added thing that you're gay, and that supposedly the tradition says you're a piece of shit and rejects you, like [Rabbi Stern's sermon] on Yom Kippur, or your mother is ashamed of you, or whatever. So you have all this negative energy coming from the Jewish world that, if it's not important to you, you just dump it. So, Jewish partner, schmewish partner, it doesn't matter, that's why."

A secular Jew, Donald, expressed his anger about the subject: "I'm very offended by the idea that I should limit my partners to only Jews, because it's hard enough as a gay man to find a partner in the first place. And even if I do, I don't necessarily get the encouragement and the support from the Jewish community!"

A traditional Jew, Nathan, reported that, as a Jewish educator, he is very familiar with the phenomenon of out-marriage in the heterosexual Jewish community. He observed that many of these Jews were marrying non-Jews, partly due to negative Jewish experiences and associations. He commented that "somebody once said the Jews are just like everybody else, just more so. It's a great line. I think gay Jews are like all the other Jews, just more so!"

Nathan makes the important point that, if it is true that some heterosexual Jews are marrying out of the faith due to negative Jewish experiences, then considering the levels of disapproval from the Jewish community toward gay Jews, it should not come as a surprise that gay Jews are partnering at an even higher rate with non-Jews.

A secular respondent suggested that once a Jewish man has already "violated" ethno-religious standards by coming out as gay, it is then less difficult to break other community norms. In his words, "Given that I got past all the stigma and that I am attracted to the same sex, is it going to be a such a big deal? If I went that far... I mean, that's a big, big, big taboo for many years. So if I got past that, and I meet someone with a real chemistry and he's not Jewish, who cares?

As mentioned, more respondents reported that their parents prefer them to partner with Jews rather than with non-Jews. There were some cases, however, where respondents reported a double standard on the part of their parents. Because some parents assumed that their gay sons would not have children, they did not express a preference that they have a Jewish partner. For their heterosexual children, however, these same parents did express a preference for a Jewish spouse. This double standard among parents has also been identified by Mushkat (1999, 242–43) in her research on Jewish lesbians. This attitude likely further accentuates the trend of inter-group partnering.

This evidence suggests that gay Jews may be dealing with conflicting desires. On the one hand, there may be a desire to find Jewish partners for reasons of cultural and ethno-religious connection. On the other hand, due to over-all non-acceptance by the Jewish community, there may be competing forces pushing gay Jews away from Jewish life and other Jews. The following sections explore other ways that community non-acceptance can directly or indirectly foster a situation of inter-group partnering.

c) Lack of Gay Jewish Infrastructure

Another reason why gay Jewish men who wish to find Jewish partners are not successful in doing so is the lack of a gay, Jewish, organizational infrastructure in Toronto to facilitate social interaction among gay Jews. Several respondents complained that there are few opportunities for gay Jewish men to meet one another. Stuart admitted, "I've always thought the next partner was going to be Jewish, and it's just never turned out that way. I guess if I can be involved more in the gay Jewish community, I would meet more gay Jews, and the opportunity [would increase], but if the community isn't there, then it's always hit and miss whether [he] turns out to be [Jewish]." Jordan also lamented that in a large Jewish center like Toronto, there are few safe Jewish settings for gay Jews to socialize. "There aren't Jewish gay socials. Kollel [a liberal Jewish educational institution in Toronto] runs this fabulous singles dating thingy. And a couple of years ago, they made it abundantly clear that it was for straight people. And a couple of us challenged them on that one and said, Why? And their answer was that they need to make it really clear that people are coming into a safe environment, and part of that safety is knowing that it is purely a heterosexual environment!"

Because Toronto is by far the largest Jewish center in Canada, one might assume that it would have many opportunities for gay Jews to meet one another. This is not the case. The largest gay Jewish organization in the city, Keshet Shalom, officially folded in the fall of 2001. Data suggest that a lack of community support, both financial and emotional, played a significant role in its demise (Schnoor 2003, 34-36). There were simply not enough openly gay Jews in town who were willing to continue to work on a purely volunteer basis to sustain the organization. There is a gay Jewish social group called Kulanu, headquartered at the University of Toronto, that meets a few times a year. However, it is not well-publicized outside of the university community and is designed for "young adults." Numerous respondents over the age of thirty-five complained about the absence of any gay Jewish programming geared to their age group in the city.[14]

As compared to the high priority the Jewish community puts on developing and funding dating and matchmaking services for heterosexual Jews, no community-endorsed service exists for matching gay Jews. One can reasonably speculate that this is directly related to the community's non-acceptance of homosexuality and gay partnerships. This phenomenon is not unique to the Jewish community. Peplau et al. (1997, 20-21) and Greene and Boyd-Franklin (1996, 260) report the same situation in Black communities, as does Chan (1989, 18-19) for Asian communities. Minority communities are reluctant to develop the necessary social infrastructure for their gay men and lesbians to meet one another.

d) *Internalized Distaste or Discomfort with Jewishness*

As mentioned earlier in this work, Jackson and Sullivan (1999, 4-6) reported the tendency in Australia for Anglo-Australian gay men to partner at high rates with Asian gay men. This pattern is caused in part by the fact that, similar to what has been described above, gay Asians feel alienated from their own culture due to the rejection of their homosexuality from their host community. In the Australian case, this type of inter-group partnership can also involve issues of exploitation and racism. Older, gay, Anglo men are stigmatized in the Australian gay scene due to their age. Therefore, they do not have the required social capital to partner with younger Anglo men. However, by virtue of their Anglo ethnicity, these older men still hold higher social and cultural capital than Asian gay men who are stigmatized due to their non-Anglo identity. Therefore, the predominant pattern sees older, gay, Anglos partnering with younger gay Asian men.

We offer this example to illustrate how racial oppression can affect choice of partner. Older gay Anglo men settle for gay Asians because this is the best they can do. To add to the devaluing of Asian culture in this context, younger gay Anglo men speak pejoratively of these older Anglo men as "rice queens" because of their tendency to partner with Asians. Interestingly, gay Asians who partner with other gay Asians are also given a pejorative term: "sticky rice." The results of these patterns can be very damaging to the psyche of the gay Asian man. When Asians are rejected by their own culture for being gay and then also suffer from the racism of members of the Anglo-culture for being Asian, internalized racism can occur (Han 2000, 219-21; Nakajima et al. 1996, 573-74). Similar patterns of internalized racism among gay Blacks in the United States are reported by Greene (1998, 49-51). This devaluing of one's own ethnicity can accentuate inter-group dating patterns.

Similar to the case of Asians, gay Jews can also feel doubly stigmatized. On the one hand, they can feel marginalized due to a perceived Christian-normative climate in the general society, while on the other, they can feel marginalized due to the rejection of their gay identity from their own Jewish community. The interview data did not indicate whether inter-faith dating patterns among gay Jews contain elements of exploitation as described above.[15] However, what did come to light was some resentment toward Judaism or a devaluing of one's Jewish identity on the part of some respondents. Sometimes this resentment manifests itself as a subconscious distaste or discomfort for things Jewish. Ten respondents acknowledged the reality of this internalized distaste for Jewishness. Evan raised the issue in the context of partnering practices. "I think that a lot of gay men will tell you, I've dated lots of Jewish men and a lot of them are fucked up. You heard me say it! [laughter]. I can't parse out how much of that is my own anti-Jewish stuff, my own issues around Jews and Judaism and my own fears around that. This has almost been the universal experience of gay men in the Jewish community: it's been so negative that they carry around all this baggage about being Jewish."

Lawrence's perspective is quite instructive in this regard. While he states a preference for a Jewish partner, he then goes on to express why he does not actually want one: "I find that I always thought it would be great to have a Jewish partner because I wouldn't have to explain so many things. But every time I have tried to meet or date Jewish guys, they always remind me of my family, and so it's a horrible situation. It's not what I want, so they're better off as friends; but the idea still appeals to me."

Other comments from respondents regarding their reservations about a Jewish partner included, "If I found out someone was a gay Jewish man, it would be unpleasant for me because people might think that I was Jewish," or "Sometimes their angst reminds me a little bit of my angst," or "I get very bored of the Jewish mother, overprotective Jewish son schtick," or "I think that men having neuroses too close to your own aren't very hot. Non-Jewish men's neuroses aren't as apparent or as oppressive." All hint at a certain discomfort with their own

Jewishness. There are a number of references to this phenomenon in the popular gay Jewish literature (Klepfisz 1989, 46-48; Mass 1994, chap. 2; Schuman 1989, 16). Rofes (1989, 202-03), a gay Jewish man, goes so far as to say that his anti-Jewishness intensifies when he dates Jews.

Other examples in the data that may have some connection to a discomfort with Jewishness concern issues of sexual and erotic attraction. A secular, gay Jew confided that he finds having sex with a Jewish man to be very unappealing. He admitted it is not rational. He referred to it as a "gut feeling." Another respondent in his fifties, Ben, admitted that "every time that I have dated Jewish men, it becomes a matter of friendship. The sexual thing for me, it just disappears. It's more than a coincidence. It's got to be more. There is almost a collegiality about meeting another Jewish gay man. There is a framework that is similar...which you think would propel you into a relationship with a Jewish man. But for me, [it doesn't]."

Glen provided an interesting commentary regarding why gay Jewish men may not find other gay Jewish men sexually arousing: "A lot of gay cultures have been based on the sort of California surfer-boy, twink thing, and it has not been a typical physical characteristic of gay Jews to be like these blond, blue-eyed, hairless children. So if that's the kind of culture you're brought up in and looking for as desirable, then you're not going to look for Jews, because they don't fit that characteristic." Because the media has glorified a certain physical type at the expense of other physical types, gay Jewish men may subconsciously internalize these preferences and develop the idea that Jews are not sexually desirable. Another respondent had a similar perspective: "Are we attracted to other Jewish men? Most would say no, because physically, they're not [our] type. Every gay Jewish man I've met is usually dark: Dark hair, dark eyes; we all want blonds; we want Brad Pitt!; we're not that interested in Adam Sandler! It's just unfortunate," and "Even that [gay] rabbi from Los Angeles we brought up here for the weekend...the knife through my heart was that his lover is a blond, blue-eyed, Catholic guy. The Rebbetzin [rabbi's 'wife'] wasn't even Jewish! I couldn't believe it!"

Steven provided another interesting observation regarding sexual expression. "Another [issue] could be sexual practices. Having read the messages for a gay Jewish group, there seems to be a lot of attitudes that gay Jews should be very Christian missionary in their sexual behaviour. So good Jewish boys aren't into S and M, aren't into anything kinky, they're just a good fuck, and that's it. So if they're into anything outside of that, they feel like they're the only gay Jew who's into something alternative, and therefore they cannot look within the gay Jewish community for it; they have to look outside." Again we see that gay Jews may internalize certain stereotypes about Jews that may restrain them from their desired form of expression, thus potentially preventing Jews from becoming sexually involved with one another.

It is useful here to briefly reflect on gender differences in partnering practices. Although the qualitative component of our study has no direct information on lesbians, two respondents reported a difference between Jewish gay men and Jewish lesbians in terms of their priorities in forming relationships. Emile reported that he was amazed to see so many Jewish lesbian couples. Another respondent, Abe, offered the following reason why there would be a difference between genders: "I think the nature of how women want to connect and how men want to connect are very different. I'm making very gross generalizations, but I think women see relationships in the long-term, and [sexual] attraction goes into a pool of other things, whereas men are just more likely to act on impulse and attraction. I think that the appeal for men tends to be very genital-brain, whereas for women, there's a different kind of culture." This insight is probably just as applicable to heterosexuals as it is to gay men and lesbians. Abe's speculation is corroborated to some extent by Hunter et al. (1998, 94), who reported that gay men, more than lesbians, are inclined to have sex on the first date. Abe's perspective also lends some support to the idea brought forward in this study that gay Jewish men seem to be interested in forming relationships with other Jewish men for cultural reasons, but sometimes fall short of doing so owing to a lack of sexual attraction. It is likely true that Jewish lesbians share this desire for cultural connection with other Jews. If it is true that lesbians put less emphasis on immediate sexual attraction in their relationships, then this may be one factor which accounts for a higher inter-group partnering rate among gay Jewish men than among Jewish lesbians. More exploration of this topic is needed before definitive conclusions can be drawn.

Whether it be a distaste for Jews as sexual partners, a distaste for overprotected Jewish sons, or a distaste for Jewish angst or neurosis, our data reveal that some respondents demonstrate a subconscious, or "gut," discomfort for things Jewish that may prevent them from establishing long-term, romantic partnerships with other Jews.

Of course, these patterns are not limited to gay Jews. Hyman (1995, 169) and Prell (1999, 18-20) have argued that in their struggle to find a secure place in American society, American (heterosexual) Jews have historically projected those qualities for which Jews have been criticized (for example, being overly demanding, pushy, money-hungry) onto other Jews as a way to try to separate themselves from this damaging stigma. This has translated into the prominent gender stereotype of the Jewish woman as a Jewish American Princess (JAP), for example. This negative stereotype, which enjoyed its greatest popularity in the late 1970s and 1980s but is still relevant today, depicts Jewish women as materialistic, whiny, aggressive, nagging—and withholding of sex (Fishman 2000, 8-9; Prell 1999, chap. 6). Jewish men thus sometimes wish to avoid these sexually passive women and seek out their dream shiksa (non-Jewish woman), often portrayed as blond and blue-eyed, who will supposedly be more sexually exciting and materialistically less demanding, as well as less Jewish looking (Weinfeld 2001, 129). In addition, Jewish women may internalize these negative stereotypes about themselves and seek out non-Jewish partners as a way of separating themselves from the environment that they associate with the oppression. This attitude, exhibiting a distaste for Jewish women, is not dissimilar to the examples above of gay Jewish men having a distaste for other Jewish men. Both can promote some form of internalized distaste of Jewishness and thus might contribute to inter-group partnerships and marriages.

e) Internalized Homophobia

There is some discussion in the gay Jewish literature of a form of internalized oppression often referred to as internalized homophobia (Duberman 1991, 14-15;

Raphael 1996, 84; Schuman 1989, 16). When a gay individual is living in a community that is quite hostile to homosexuality (as the Jewish community can be), this person may start to internalize these negative messages and develop shame and guilt about his/her homosexuality. This phenomenon can also contribute to inter-group partnering on the part of gay Jewish men. While ten respondents demonstrated attitudes that may be connected to internalized distaste or discomfort with Jewishness, only two interview respondents demonstrated some form of internalized homophobia.

The most prominent example of this phenomenon came from a respondent who described his feelings of guilt over being gay. His guilt developed because he believed that in being gay, he was shirking his responsibility to fulfill the requirement in the Jewish community to procreate to ensure Jewish continuity. Interestingly, this attitude had a strong bearing on his choice of partner. "I felt guilty being Jewish and gay myself. It was a big source of discomfort for me. I would not have entertained a Jewish gay guy, because it was enough that I'm Jewish and gay, so therefore that's one less male in the [Jewish] stable, so to speak, likely to procreate. But the idea that I'd be – even though I obviously wasn't going to make the other guy gay – the idea that I'd be taking another gay guy, that I'd be with another gay guy, was just too much for me to deal with. I just couldn't accept that." This respondent felt shame for being gay within a Jewish context and seemed to have felt more shame about the idea of "contaminating" another gay Jew by partaking in a relationship with him. This internalized homophobia prevented him from seeking Jewish partners.

Conclusion

This study contributes to the understudied topic of the ethno-religious composition of minority gay male relationships. It illustrates that while the majority of gay Jewish men may desire a Jewish partner, there are a number of reasons why these partnerships do not actually occur. Neither the type of Jewishness nor age proved to be a pivotal factor. In our sample, we found both traditional and secular Jews of different age ranges with non-Jewish partners. The fact that so many gay Jews are partnering with non-Jews does offer some evidence to suggest that anti-Jewish sentiment in the gay community is not widespread.[16]

The study pieces together different aspects of the gay Jewish experience and illustrates the links to the larger phenomenon of inter-group partnering practices. In addition to the more obvious factors at work, such as the small number of gay Jews, more subtle underlying dynamics are outlined. The theme of non-acceptance of homosexuality by the Jewish community can sometimes create a situation where gay Jews are turned away from Judaism and Jewish life and find it more comforting to seek a partner outside of this oppressing milieu. The fact that gay Jews often feel pressured to stay closeted, as well as the Jewish community choosing not to develop a social infrastructure to allow its gay members to meet one another, accentuates the situation. Our findings suggest that the perception of generalized, anti-Jewish sentiment combined with the rejection of gay identity in the Jewish community can engender internalized oppression on the part of gay Jews.

Expressions of internalized distaste or discomfort with Jewishness or internalized homophobia can contribute to inter-group partnering factors. Other minority communities, such as Blacks and Asians, demonstrate similar patterns.

It is important to note that this study dealt with the question of whether the partners of our respondents were Jewish or non-Jewish at the time the relationships began. Sociologists of the Jews (for example, Fishman 2002; Phillips 1997) are beginning to pay closer attention to how Jewish and non-Jewish partners negotiate their mixed identities during the course of their relationships. Interestingly, among the eight respondents who had non-Jewish partners, one reported his partner is about to begin the process of converting to Judaism, another reported that his partner is considering converting to Judaism, and two others reported that their partners are "interested" or "knowledgeable" about Judaism. Perhaps what we are seeing here are gay Jewish men who because they cannot find other gay Jewish men (due to small numbers or lack of social infrastructure), are therefore partnering with the next best thing: non-Jewish gay men who have a favourable predisposition to Judaism. Another explanation is that these Jewish men are in fact influencing their partners during the course of their relationships to acquire an interest in Judaism. If the latter explanation is true, then gay Jewish men, as is sometimes the case among heterosexual Jews, may actually be helping to bring more Jews into the fold. Toronto's more traditional Jewish nature should also be taken into account here. Future research is recommended to determine the rate at which non-Jewish partners of gay Jewish men are converting to Judaism. Based on the 1990 NJPS, we know this rate is approximately ten to twenty percent in the United States for the non-Jewish partners of heterosexual Jews (Cohen 1991, 49).

Overall, the study provides an example of the way some gay Jews suffer from what could be termed "identity ambivalence." This ambivalence is caused by conflicting or competing desires: on the one hand, they want Jewish partners for cultural reasons, but on the other, there are forces at work that do not allow this to happen. Respondents often experience a clash between their Jewish and their gay desires, and they are forced to struggle to find ways to negotiate this difficult divide. These circumstances make it difficult for gay Jews to develop personally meaningful, Jewish identities which integrate their gayness. It appears that having a relationship with a non-Jewish partner who is fond of Jews and Judaism may be one strategy that helps to bridge this divide. Future research might explore other strategies of identity negotiation for gay Jews. Research on partnering in other gay and lesbian minority groups can explore whether a comparable identity ambivalence exists there, as well. In addition, the issue of whether gay fatherhood may play a role in terms of the desired ethnicity of the partner requires further exploration.

ACKNOWLEDGEMENTS

The research reported here was supported by the Memorial Foundation for Jewish Culture. The authors would like to thank Professor Leo Davids of York University for facilitating the purchase of 2001 Census Canada. Ruth Sutcliffe of Statistics Canada was particularly helpful in identifying the precise Census data needed for the specific concerns of this research project.

The authors would also like to thank the two anonymous *Canadian Ethnic Studies/Études ethniques au Canada* reviewers for their helpful comments.

BIOGRAPHIC NOTES

Randal F. Schnoor teaches at York University in the Division of Social Science. He holds a Ph.D in sociology from McGill University. He specializes in the sociology of contemporary Jewish life in North America. Among his recent publications is an article on Hasidic Jews in Montreal published in *Canadian Jewish Studies*, 2002. A forthcoming monograph will further explore strategies of identity negotiation among gay Jewish men.

Email: randal.schnoor@primus.ca

Morton Weinfeld is a professor in the Department of Sociology at McGill University, where he holds the Chair in Canadian Ethnic Studies. Among his recent publications is the 2001 volume *Like Everyone Else...But Different: The Paradoxical Success of Canadian Jews* (Toronto, McClelland & Stewart). He is currently working on a study of cultural sensitivity and the childbirth experience of immigrant women in Montreal.

Email: morton.weinfeld@mcgill.ca

NOTES

1. The 2000 National Jewish Population Study reported a slightly higher finding than 1990. Calculation methods differed, however.

2. Unless otherwise stated, for this study, Jews in Canada are calculated by combining those who identify as Jewish by religion with those who identify as Jewish by ethnicity (with no religion reported).

3. Reference persons correspond to "Person #1" on the Census questionnaire. If the Jewish member of the same-sex common-law relationship was "Person #2" on the Census questionnaire, he/she would not be included in this calculation. As far as we can ascertain, custom tables purchased from Statistics Canada only allow such calculations to originate from "Person #1."

4. This rate of 89 percent applies when classifying the partner of the Jewish person by ethnic origin. When classifying the partner of the Jewish person by religion, very similar rates are found. Of the 700 reference persons who are Jewish (by religion) and in a same-sex, common-law relationship, only 105 partnered with a fellow Jew. Thus, 595 out of 700 gay and lesbian Jews did not. This corresponds to an out-marriage rate of 85 percent.

5. The figure of 89 percent should be treated with some caution. It may not be directly comparable to the 30 percent rate. Whereas the out-marriage rates of heterosexual Jews include Jews from the entire spectrum of very secular to very religious, this may not be the case with the out-marriage rates of gay and lesbian Jews. More observant or traditional Jews would be less likely to embrace a gay identity, and to have a same-sex partner and feel comfortable indicating this on the census. Thus, those Jews who contribute to the 89 percent inter-marriage rate are likely to be disproportionately secular.

6. By traditional Jews, we include both Jews who are religiously-observant, Orthodox Jews (there were three such Jews in our sample), as well as those Jews who might not observe all Jewish laws but still maintain a conservative religious orientation and place more importance on synagogue attendance, celebrating the Jewish holidays, and preserving and perpetuating Jewish traditions than do liberal Jews, who are quite secular in their orientation, attending synagogue and celebrating Jewish holidays infrequently.

7. Shaffir and Weinfeld (1981), 13 report that Canadian Jews "speak more Yiddish, provide their children with more intensive Jewish education, make higher per-capita contributions and relatively more visits to Israel, are more likely to be Orthodox and less likely to be Reform, and have lower rates of intermarriage." Brodbar-Nemzer et al. (1993), 43 reported that based on 1990 survey data, 19 percent of Canadian Jews identify as Orthodox, while only 9 percent of American Jews do so; 43 percent of American Jews identify as Reform, while only 11 percent of Canadian Jews do so. They also reported that on a variety of Jewish ritual behavioural items (for example, lighting Sabbath candles, attending Passover seders), Canadian Jews consistently observe at a higher rate.

8. The interview questionnaires are available from the first author upon request.

9. This discussion on partnering should not be interpreted as advocating long-term monogamous relationships as the "ideal" or "proper" form of personal relationship. The authors are well aware that other forms of relationship models, such as more casual, non-monogamous physical relationships with a number of partners, are common among gay men. This being the case, it is interesting to note that the vast majority of respondents in this study did express a desire to enter into a long-term monogamous relationship if the opportunity presented itself.

10. Weinfeld found that Jews were more likely to identify themselves in personal ads by their ethnicity than other ethnic groups. They were also more likely to specify that they are seeking partners with the same Jewish ethnicity.

11. The interviews primarily gathered data on respondents who were involved in a serious relationship (that is, with a single partner or boyfriend). In retrospect, it would have been useful to collect more data about the ethnicity of the men respondents were casually dating.

12. Of course, it is possible that a small proportion of the married Jewish men included in the NJPS are gay.

13. This figure is taken from a 1998 website entitled "Partners Task Force for Gay and Lesbian Couples: Are Gay Couples Overlooked?" (Partners' Gay/Lesbian Media Survey) [http://www.buddybuddy.com/toc.html].

14. In 2002 a Jewish Gay and Lesbian Fund was developed in the Toronto Jewish community to sponsor gay Jewish cultural and educational programs in the city.

15. The one exception to this is the case of a respondent who reported that he had a non-Jewish partner who "liked being humiliated in sexual activities." When the relationship ended, "he finally let me know that the reason he wanted me as a sexual partner was because I was Jewish, and he could think of nothing more humiliating than to have sex with a Jew." The respondent also reported that some of his partners have fetishized him as a Jew, enjoying sex with him because they believed that Jews have bigger penises and are more sexually exciting.

16. It is also possible that some non-Jewish gay men choose Jewish (sexual) partners because they have exoticized or "othered" them. Our data did reveal one such case. See note #15.

REFERENCES

Ambert, Anne-Marie. *Same-sex couples and same-sex-parent families: Relationships, parenting, and issues of marriage*. Toronto: Vanier Institute of the Family, 2003.

Bell, Alan P., and Martin. S Weinberg. *Homosexualities: A study of diversity among men and women*. New York: Simon and Schuster, 1978.

Brodbar-Nemzer, Jay, Stephen M. Cohen, Allan Reitzes, Charles Shahar, and Gary Tobin. An overview of the Canadian Jewish community. In *The Jews in Canada*, ed. R.J. Brym, W. Shaffir, and M. Weinfeld, 39–72. Toronto: Oxford University Press, 1993.

Chan, Connie. 1989. Issues of identity development among Asian American lesbians and gay men. *Journal of Counselling and Development* 68: 16–20.

Cohen, Steven M. *American assimilation or Jewish revival?* Bloomington: Indiana University Press, 1988.

———. *Content or continuity?: Alternative bases for commitment*. New York: American Jewish Committee, 1991.

Davids, Leo. Tracking demographic assimilation. In *Jewish survival*, ed. Ernest Krausz and Gitta Tulea, 207–26. New Brunswick, NJ: Transaction, 1998.

Duberman, Martin. *Cures: A gay man's odyssey*. New York: Plume, 1991.

Dworkin, Sari H. Female, lesbian and Jewish: Complex and invisible. In *Ethnic and cultural diversity among lesbians and gay men*, ed. B. Greene, 63–87. London: Sage, 1997.

Fishman, Sylvia Barack. *A breath of life: Feminism in the American Jewish community*. New York: Free Press, 1993.

———. *Jewish life and American culture*. Albany: State University of New York Press, 2000.

———. *Relatively speaking: Constructing Jewish identity in Jewish and mixed married families*. Ann Arbor: Frankel Centre for Judaic Studies, University of Michigan, 2002.

Goldstein, Sidney. Highlights of the 1990 national Jewish population study. *American Jewish Yearbook*, 77–173, 1992.

Greene, Beverley. Lesbian couples. In *Dyke life: From growing up to growing old—A celebration of lesbian experience*, ed. K. Jay, 97–106. New York: Basic, 1995.

———. Family, ethnic identity, and sexual orientation: African-American lesbians and gay men. In *Lesbian, gay, and bisexual identities in families: Psychological perspectives*, ed. C.J. Patterson and A. D'Augelli, 40–52. London: Oxford University Press, 1998.

Greene, Beverley, and N. Boyd-Franklin. African-American lesbians: Issues in couples therapy. In *Lesbians and gays in couples and families: A handbook for therapists*, ed. J. Laird and R.J. Green, 251–71. San Francisco: Jossey-Bass, 1996.

Han, Shinhee. Asian-American gay men's (dis)claim on masculinity. In *Gay masculinities*, ed. P. Nardi, 206–23. Thousand Oaks, CA: Sage, 2000.

Hunter, Ski, Coleen Shannon, Jo Knox, and James I. Martin. *Lesbian, gay and bisexual youth and adults: Knowledge for human services practices*. Thousand Oaks, CA: Sage, 1998.

Hyman, Paula E. *Gender and assimilation in modern Jewish history: The roles and representations of women*. Seattle: University of Washington Press, 1995.

Jackson, Peter A., and Gerard Sullivan. *Multicultural queer: Australian narratives*. New York: Haworth, 1999.

Klepfisz, Irena. Anti-Semitism in the lesbian feminist movement. In *Nice Jewish girls: A lesbian anthology*, ed. E. Torton Beck, 51–60. Boston: Beacon, 1989.

Mass, Lawrence D. *Confessions of a Jewish Wagnerite: On being gay and Jewish*. London: Cassell, 1994.

Mushkat, Danita. "Alienated Jews": What about outreach to Jewish lesbians? *Journal of Jewish Communal Service* 75, 4 (1999): 239-47.

Nakajima, Gene A., Y.H. Chan, and K. Lee. Mental health issues for gay and lesbian Asian Americans. In *Textbook of homosexuality and mental health*, ed. T.S. Stein, 563-81. Washington DC: American Psychiatric Press, 1996.

Peplau, Anne, Susan D. Cochran, and Vickie M. Mays. A national survey of the intimate relationships of African-American lesbians and gay men: A look at commitment, satisfaction, sexual behaviour, and HIV disease. In *Ethnic and cultural diversity among lesbians and gay men*, ed. B. Greene, 11-38. London: Sage, 1997.

Phillips, Bruce A. *Re-examining intermarriage: Trends, textures, strategies.* New York: American Jewish Committee and the Susan and David Wilstein Institute of Jewish Policy Studies, 1997.

Prell, Riv-Ellen. *Fighting to become Americans: Jews, gender, and the anxiety of assimilation.* Boston: Beacon, 1999.

Raphael, Lev. *Journeys and arrivals: On being gay and Jewish.* Boston: Faber and Faber, 1996.

Rofes, Eric. E. Living as all of who I am: Being Jewish in the lesbian/gay community. In *Twice blessed: On being lesbian or gay and Jewish*, ed. C. Balka and A. Rose, 198-206. Boston: Beacon, 1989.

Schnoor, Randal F. Finding one's place: Ethnic identity construction among gay Jewish men. Ph.D. Diss. Department of Sociology, McGill University, Montreal, 2003.

Schuman, Burt E. Confessions of a "Feygele-Boichik." In *Twice blessed: On being lesbian or gay and Jewish*, ed. C. Balka and A. Rose, 12-20. Boston: Beacon, 1989.

Shaffir, William, and Morton Weinfeld. Canada and the Jews: An introduction. In *The Canadian Jewish mosaic*, ed. M. Weinfeld, W. Shaffir, and I. Cotler, 7-20. Toronto: John Wiley and Sons, 1981.

Sharot, Stephen. Judaism and Jewish ethnicity: Changing interrelationships and differentiations in the diaspora and Israel. *Sociological Papers* 6 (1998): 87-105.

Sherif, Bahira. The ambiguity of boundaries in the fieldwork experience: Establishing rapport and negotiating insider/outsider status. *Qualitative Inquiry* 7, 4 (2001): 436-47.

Statistics Canada. Profile of Canadian families and households: Diversification continues. (Catalogue No. 96F0030XIE2001003), 2002.

———. All persons by selected religion and ethnic origins, gender and living arrangements, showing selected characteristics – for Canada and selected census metropolitan areas – 2001 Census (20% sample data). (Table No. E741tbA), 2004a.

———. All persons by selected religion and ethnic origins, showing ethnic origins and religions of same sex common law partners – for Canada and selected census metropolitan areas (Table No. E741tbB), 2004b.

Weinfeld, Morton. *Like everyone else... but different: The paradoxical success of Canadian Jews.* Toronto: McClelland & Stewart, 2001.

Negotiating Social Stigma Among Gay Asian Men

In recent years, there has been a proliferation of literature in gay Asian criticism that theorizes the experience of gay Asian men and Asian men who have sex with men in the West. In particular, critics and theorists of gay Asian experience center their analysis on the negative portrayal of gay Asian men (as passive, submissive, effeminate and less sexually attractive) in the West and its effects on these men.

This portrayal, together with the western idea of male beauty, as the critics and theorists argue, has rendered Asian men not only invisible in the mainstream gay community, but also undesirable as potential partners.

Consequently, many gay Asian men have internalized negative feelings about themselves and other gay Asian men while idealizing white men as potential partners. They see each other as competitors for the attention of the (few) white men who are interested in Asian men. Some even internalize and adhere to the stereotypes as being passive and submissive to please and attract white men (see, for example, Ayres, 1999; Chuang, 1999; Han, 2005; Kong, 2002; Nemoto et al., 2003).

Other critics and theorists of gay Asian experience argue that the stereotypes have also contributed to the sexual objectification of Asian men by some white men who, writes Chong-Suk Han, desire them 'purely for their exotic eroticism' – a fantasy that is maintained by the stereotypes. What these men 'are often attracted to in Asian men', Han goes on to write, 'is an idealized notion of a passive, docile, submissive – in short, a feminized – lover, eager to please his virile white man' (2005: 36).

Unsurprisingly, within this social context, many gay Asian men suffer low self-esteem and self-worth. To combat the stereotypes, some critics suggest increasing the visibility of Asian role models in the gay community (see, for example, Mao et al., 2002; Nemoto et al., 2003; Poon and Ho, 2002). Others propose countering 'the effects of stereotypes through developing stronger conceptions of self-identity and feelings of power in their relationships', that is, choosing 'to date' exclusively API [Asian and Pacific Islander] men to avoid the racial power dynamics with White partners' (Nemoto et al., 2003:13; see also Drummond, 2005). Still others suggest that, in addition to supportive families and close friends, involvement with the mainstream gay community can help Asian men 'to cultivate a positive and comfortable self-identity' (Mao et al., 2002: 428).

In contrast, we explore resistance to the stereotypes at a personal level – a topic that we think is often overlooked in gay Asian criticism. Using narratives obtained through a series of interviews, we analyze the ways in which they negotiate and (re)frame the social stigma associated to their (Asian) bodies and erotic desire for white men. 'Framing', as Martin Bloom and associates write, 'refers to how a person constructs his or her psycho-social reality' (1991: 531). Language, as Ronald Adler and George Rodman remind us, does not simply 'describe the world' but also, and perhaps more importantly, shapes 'the way we look at the world and, in so doing', influences 'the way we behave' (1991: 65). Changing our ways of talking, then, will also and inevitably change the way we construct our social world.

Methodology

The primary material we use in this article was drawn from a community-based study that we conducted in 2001. The purpose of the study was to develop a culturally appropriate HIV prevention program for men of East and Southeast Asian origin who use internet gay chatrooms in Toronto. However, in order to do so, we believed that it is essential to understand how these men function within the socio-economical context in which they are located as it inevitably affects their understanding and behaviours about HIV/AIDS. Interviewing, which allows participants to articulate their psychosocial experience in their own terms, was selected as the data collection method. Original data analysis aimed at understanding this context and how it relates to HIV/AIDS in this population (see Poon et al., 2005).

During our data analysis, it also became quickly apparent to us that social stigma is one of the central themes expressed by the participants. Many discussed quite extensively their experience of the negative stereotypes associated with being gay and Asian and yet, in spite of that social stigma, they (re)framed their bodies and their desire for white men. We thus performed a sub-set of analysis and, in this article, explore these meanings.

Study participants were recruited from gay.com (Toronto-Asian chatroom) in which we advertised the study, provided information about our agency, and the study goals and methodology, and attempted to develop trust and rapport with potential participants. To obtain a group of diverse participants that allowed us to compare experiences of different Asian MSM chatroom users and thus refine concepts, our recruitment was accomplished at different times of the day, as we assumed that people who used the chatroom at different times of the day might represent different types of users (e.g. older users may use the chatroom more often during the evening and at night simply because they need to work during the day; in contrast, younger users may use it more often during the day because their parents are home during the evening and at night). At first, study participants recruited tended to be gay, younger (under 30 years old) and Chinese; we then specifically aimed at recruiting users who did not identify themselves gay, and who were older (over 30 years old) and not necessarily Chinese, through word of mouth in the chatroom. The main criteria for participation in this study were being a male of East and Southeast Asian descent and using internet gay chatrooms in Toronto.

Study participants (n = 21) were asked to self-identify themselves ethnically; eight said they were Chinese, one Chinese-Korean, one Chinese-Japanese, one Chinese-Vietnamese-Thai, one Hong-Kong-Chinese, one Korean, two Filipino, one Taiwanese, one Spanish-Filipino, three Vietnamese and one Vietnamese-Canadian. However, five were born in Canada, three in China, three in Hong Kong, one in Korea, one in Laos, two in the Philippines, two in Taiwan, one in Thailand and three

247

in Vietnam. About half the study participants (n = 10) had lived in Canada for about 11–20 years, another five for over 20 years, and the remaining five for less than 10 years (one unknown). Except for one who self-identified himself as bisexual, the rest said that they were gay. Their age ranged from 18 to 39 with a mean of 26.1. The study participants were highly educated, except for one who had finished only high school; the rest had some level of undergraduate education or above (17 undergraduate education and three postgraduate education). There was a fairly wide range of income. Eight participants were earning less than $10,000 per year, five were earning between $20,001 and $30,000, one earning between $30,001 and $40,000, two earning between $40,001 and $50,000, one earning from $50,001 and $60,000 and four were earning more than $70,001. Although study participants were selected to represent a wide diversity in ethno-cultural background, age, social class and educational level, narratives obtained in this study cannot be generalized to the Asian MSM population as a whole.

All interviews were conducted online through real-time chat, 'in which messages are written and read at the same time, though in different places' (Mann and Stewart, 2000: 11). Some critics suggest that this method – online interviewing through real time chat – is impersonal and distancing; thus, it cannot facilitate detailed exploration of meaning, narrative and discourse (Davis et al., 2003). However, other researchers challenge this assumption and argue that the quality of data depends on the rapport and shared research agenda between the researchers and participants rather than the medium itself – particularly if participants find the research topic relevant to them and feel that they are given an opportunity to be 'heard' in a meaningful way. Thus, highly interactive forms of communication can be developed online (see Mann and Stewart, 2000; Eysenbach and Wyatt, 2002).

Based on our personal and professional experience, we knew that in many Asian cultures, sex and sexuality are a very sensitive topic and are rarely permitted to be discussed openly. Because of the different physical spaces occupied by the researchers and participants, online interviews might enable participants to discuss sensitive issues without feeling embarrassed or being judged. Moreover, participants are interviewed in places where they feel comfortable, 'which might allow communication to flow with a minimum of self-consciousness and self-restraint . . . As a consequence, interaction in the virtual field may lead to more direct communication and greater self-disclosure even about sensitive issues' (Mann and Stewart, 2000: 200). Furthermore, we recruited participants through Gay.com. As users, they were already familiar with the chat format prior to the interview. For these reasons, we decided to interview participants online.

To ensure confidentiality, all participants were interviewed in private chatrooms within Gay.com. The interviews were conducted by one of our research team members who identified himself as gay and Asian. He was familiar with the local Asian MSM community and the issues they faced; thus, it was expected that his personal associations and knowledge would make it easier for him to connect with the study participants and help facilitate the interviews. Typically, interviews lasted about two hours. All participation was voluntary and no gift or honorarium was offered. To ensure that participants met the criteria for participation, we asked each of them to ethnically identify himself and, if available, read his personal profile on Gay.com before the interview proceeded.

Through a series of unstructured, open-ended questions, participants were asked to describe their experiences of internet gay chatrooms (such as the social stigma of being Asian and gay, dating and relationships) and how they deal with the social marginalization that they experienced in the gay community. In so doing, we enabled the study participants to discuss the issues most relevant to them while attributing meanings and interpretations to events and relationships in their own terms.

The interviews were recorded within Gay.com and reformatted into Word documents. Data were generated online through real-time chat, in which participants described their experiences of chatrooms through a mutual dialogue with the interviewer; thus, we treated the data as a 'dialogue' between the interviewer and participants. As Chris Mann and Fiona Stewart write, 'people do not "write" to their friends on the internet, they "talk"' (2000: 182). However, we acknowledged that this new form of communication (talking in writing) poses certain challenges.[1]

Guided by a *constructivist grounded theory*, the analysis sought to identify 'repeated patterns of happenings, events or actions/interactions that represent what people do or say, alone or together, in response to the problems and situations in which they find themselves' (Strauss and Corbin 1998: 130). Carefully reading the transcripts several times, we first conducted line-by-line open coding to highlight key phrases that represent themes. We then developed a list of emerging themes and categories by grouping similar coded phrases together. To further refine themes and create clusters, we grouped similar themes together. Lastly, by comparing themes against one another, we developed categories and examined characteristics and dimensions of each of the categories identified. This process was performed independently by each of the authors. Subsequently, we met to discuss and compare the coded transcripts to determine, through mutual consensus, core categories and how these categories related to each other. All participants' names used in this article are pseudonyms.

Negative stereotypes

The representation of gay Asian men in the West has long been an area of study in gay Asian criticism (see Cho, 1998; Eng and Hom, 1998; Leong, 1996). For example, using gay video pornography as a site of analysis, Richard Fung illustrates the ways in which the West portrays gay Asian men and maintains its cultural hegemony through representing them as the 'Other'. As Fung argues, in gay pornography, Asian men are depicted not only as asexual, but also as submissive and are 'always in the role of servant' (1996: 189). Such images, as told by our participants, are widespread in the gay community. For example, when referring to the stereotype commonly associated with Asian men in the gay community, our participant, Will, said, 'I think we're always expected to be bottom'. Like Will, Jack stated, 'The stereotype that I think is very common is some skinny, short, Asian persons dressing in tight clothes and this image remains in everyone's head on-line'.

Characterized by 'shared beliefs – a set of characteristics believed by large numbers of one group to be true of another', as Carl James purports, a stereotypes does not affect us in abstract terms (1999: 138). These shared beliefs and assumptions are not merely viewpoints or attitudes. They are a form of social knowledge or, in Anthony Giddens's terms, 'practical consciousness knowledge' that constitutes our social worlds through social interaction:

> What enables social actors to reproduce structure is their knowledge of social life. This knowledge is both the knowledge which they use to order their own actions and make sense of the actions of others in an interactive context. Because social life is so complex, the knowledge of structure is largely of a tacit and taken-for-granted nature. When actors speak a language fluently, interact competently or see reality in a meaning-given familiar way, the knowledge which enables them to do so is of a practical, unevaluated, taken-for-granted, character. (Haugaard, 2002: 148)

A stereotype then is not simply a belief or idea about a person or a social group but also, and perhaps more importantly, structures the ways in which we perceive and interact with each other. It is (re)enacted, though unconsciously, in and through our everyday social interactions (Isajiw, 1999: 144–7; Witter, 2003).

For example, Brian spoke in anger about his experience in the gay community in which people, as a result of the stereotype, commonly expected him to be subservient:

> People think you're stupid or something or that the Asian race is a very inferior race. I find that this village is full of white supremacists. I think that they expect complete subservience and that they expect cheap sex or something from our

female portrayals or something through Asian portrayals on TV that the females are subservient and the males all know kung fu.

Similarly, Justin found in the gay chatrooms that, due to the stereotype, he was commonly treated as if he was socially inferior and thought of as a new immigrant who speaks little or no English even though he was born in Canada and has lived here all his life:

> In the Asian rooms you see tons of 40 plus [white men] ... looking for younger. I found they are generally extremely condescending. They talk to me like I'm 12 ... I found in general they're looking for a passive younger 'boytoy' and they all think I speak Chinese for some reason ... It is SO bloody annoying ... and they treat you like every Asian person is fresh off the boat.

Negative stereotypes are thus a form of social violence carried out, though unwittingly, through social interaction that we continually experience in the course of everyday life.

Resistance

However, we must not assume that gay Asian men are simply being oppressed without any agency or resistance. As Giddens argues, our social world is (re)produced in and through the everyday practice of social interaction in which we are actively, though unconsciously, involved. Thus, agency and resistance always exist in us; even the most 'marginalized' groups in our society, though obviously at a disadvantage, are not completely powerless or, in Jeffrey Isaac's view, do not 'simply respond to the behavior of the powerful. The reproduction of the relationship always involves their agency, which can be mobilized as well as transform the relationship itself' (1992: 49; see also Haugaard, 2002; Sweely 1999).

Arguing that gay Asian men have agency and resist oppression does not deny the existence of structural inequality that puts certain groups in a subordinate position or implies that every one of us has equal access to social resources. It rather refuses to treat power as something that one has or has not, insisting that power circulates and is always in coexistence with agency and resistance. As Stuart Hall writes, reading through Michel Foucault's concept of power: 'Everyone – the powerful and the powerless – is caught up, *though not on equal terms*, in power's circulation. No one – neither its apparent victims nor its agents – can stand wholly outside its field of operation' (2001: 340; original emphasis). Of course, gay Asian men can never freely negotiate power, and the ways in which they resist oppression are always bound by their socio-economical characters such as ethnicity, class and sexual orientation. But, within these constraints, they continuously (re)position themselves against the hardship of oppression and seek control over their life (see Davis, 2002; Scott, 2001).

251

This positioning, as Didi Khayatt argues, is not only mediated through categories such as ethnicity, class, and sexuality but also, and perhaps more importantly, is situational and subject to change. To demonstrate this point, Khayatt writes from her own experience as a lesbian of color:

> When asked how I identify myself, who I am and where I came from, my responses vary: (a) according to the questioner (who is asking me, what I perceive is her/his interest in knowing, how she/he is going to use the answers and how I think they will be used, what the relations of power are between the questioner and myself); (b) the context of the questions (an interview, the topic at hand, the discussion which frames the question) and (c) the circumstances under which the questions are asked (friendship, intimacy, making acquaintances, first meeting, family, etc.). In each case, my answer will be different and the differences will be generated from the relative safety of the situation and the interest of the questioner. My answer will contain elements of pride, uncertainty, political correctness, concealment of aspects that can be misconstrued, or that are threatening; in the same way, it will include aspects of my social self that I feel will make my listener think the best of me – at least for a moment. (1994: 82)

Stewart Clegg also reminds us that resistance is not necessarily antagonistic or even confrontational: 'It might be thought that, in the absence of an overt conflict, there will be no resistance to power. This would be to confuse the notion of resistance per se with a particularly dramatic expression of it' (2002: 258). Like Clegg, Steve Pile, in his article 'Opposition, Political Identities and Spaces of Resistance', remarks: 'There are tiny micro-movements of resistance, barely perceptible, even invisible or covert – quiet stealthy masquerades resistant to categorisation and definition'. Later in the article, he further elaborates: 'Resistance may take place as a reaction against unfairness and injustice, as a desire to survive intolerable conditions, but it may also involve a sense of remembering and of dreaming of something better' (1997: 29–30). In other words, resistance does not necessarily involve political struggle, violence or revolution. It can 'take place in a myriad of disorganized and spontaneous ways on a daily basis' (Westwood, 2002: 135).

Hence, as Pile argues, resistance is never static; it is far more fragile and ambiguous, and is always subject to transformation: of course, the ways in which one resists often depends on the outcome of struggles but is always bound by their social conditions that make certain forms of resistance either possible or impossible. Thus, 'acts of resistance have to be understood not only in terms of their locations in power relations but also through their intended and received meanings' with reference to their social conditions and rules at the time (1997: 26). But, such understanding, as Clegg argues, must be considered interpretive:

> When intentions are articulated, they can only be so through whatever forms of discourse are socially available regarding what intentions can sensibly be taken to be, in the forms of language, reasoning and accounting for action. In this respect, talk about intentions that others might have is a reference less to their interior mental states as causal springs of putative action than to currently 'fixed' representations for making sense of what people do. (2002: 261)

In other words, our understanding of the ways in which gay Asian men resist is neither objective nor neutral, but inevitably interpretative and discursively constituted.

However, it must be noted, resistance is not necessarily always progressive or liberating. Our actions, largely governed by taken-for-granted knowledge such as common sense and norms, are always bound by unacknowledged conditions and unintended consequences that lie outside our self-understanding (see Davis, 2002; Haugaard, 2002). Thus, our everyday actions may have contributed to our maintaining and reproducing oppression without knowing ourselves as agents of oppression. As Nancy Hirschmann writes: 'Systems of power, privilege, and oppression replicate themselves through the daily and apparently innocent actions of well-intentioned people who may not be aware of the social significance of their actions' (2003: 84; see also Young, 1992: 180). In other words, 'resistance in one direction can be oppression in another' (Pile, 1997: 26).

In the following analysis, using our participants' narratives, we demonstrate the ways in which gay Asian men negotiate social stigma associated with their bodies (often seen as passive, submissive and effeminate) and their preference for white men (often seen as an indication of internalized racism). In particular, we explore how these men (re)position themselves against these stigmatizations and (re)frame them. If one does not assume that resistance is necessarily confrontational or involves violence or political struggle, as discussed earlier, it can be argued – though the participants did not directly use the word 'resistance' when speaking of their experiences – that by (re)positioning themselves against social stigma and (re)framing it, they not only defend themselves against it, but also (re)articulate who they are, which is in itself a form of resistance (see Hoffman, 1999; Manki, 2003; Simons, 1995 for the current debate on the limitations of the power/resistance paradigm).

Negotiating social stigma associated with their bodies

Only by comparison or, as Hall writes, when reading 'in context, against or in connection with one another', can images 'gain in meaning' (2001: 328). Thus, the image of gay Asian men as skinny, effeminate, passive and submissive does not carry any meaning or 'signify' on its own, but can only come into existence in relation to what it is considered not to be – the white, tall, masculine body that is not only socially idealized, but also commonly seen as desirable in the West. This image of male bodies stands in stark contrast to that of gay Asian men who are often rendered invisible and undesirable as sexual or long-term partners in the mainstream gay community. As Peter Jackson bluntly puts it: 'Within the dominant Caucasian-focused gay sexual ideology, Asian homosexual men are simply "not worth a fuck"' (2000: 184).

Unsurprisingly, within this social context, Asian men are put in a disadvantageous position when it comes to dating. For example, Jack found people in the gay community generally not interested in Asian men; consequently, he was frequently ignored, which he admitted had affected his self-esteem greatly:

> [In the] gay community online I think people will less likely chat with me when they find out I am Asian. I don't think I have an equal chance compared to someone who is Caucasian. It's just the way life is . . . offline in person people only see me, they won't talk to me so it's just as bad . . . because it's like you're Asian . . . don't talk to me. It's just the way they are and the way I am . . . cuz you get ignored and people are so rude to you . . . like at clubs you get like dirty looks and pushed around. You get this feeling of 'not belonging'.

Like Jack, Michael spoke of his experience in which gay chatroom users refused to chat with him as soon as they found out he was Asian:

> I find there's still a stigma of being Asian . . . there are guys who like Asian but they're definitely not the majority. If I message a guy and he asks if I'm Asian and I say yes, then most of the time they leave the chat. I think many guys don't find Asians attractive/sexy/masculine.

Interestingly, however, most of our participants, even though they felt somewhat badly about being rejected, did not attribute these social injuries to themselves personally. Instead, they thought of the rejections as arising from a difference in personal taste. Ted, by normalizing the existence of racism in our society, believed that he was rejected by (white) men not because he was ugly or because there was something wrong with him but simply because he was not to their taste: 'Racism exists in each and every corner, every culture . . .' people like different things . . . let them be'. In a similar manner, Ben attributed his experiences of social rejection to a difference in 'personal preference' and in turn asked: 'Why does [someone] like men more than women? There's no reason, right!'

By reframing such rejections as arising from a difference in personal taste or preference, suggestive of the biological and unchangeable nature of human beings, they changed not only their perception about the events but also, and perhaps more importantly, their negative feelings about themselves that resulted from these social injuries. For example, Emmett stated, 'I know for some people I am not their type and that's totally cool.

Everyone is attracted to something different ... that's what makes us individuals'. Likewise, soon after Michael realized that we all have preferences, he said he was no longer upset at being rejected: 'I realize that like me, everyone else has preferences ... I was a little mad at first ... it is like racism but then I realize we all have different tastes. It doesn't bother me now'.

For most of our participants, the meaning of their bodies was not determined, though indisputably affected, by the stereotypes. They did not see being gay and Asian as entirely negative even though some admitted to having negative feelings about themselves and other gay Asian men as a result of the stereotypes. For example, even Jack, who admitted to not feeling good about being gay and Asian as a result of the negative stereotypes and found 'the way many Asian people present themselves very unattractive', insisted that Asian people had many great qualities: 'Let me tell you Asian people have the best personalities cuz my gay Asian friends are very nice people. Gay Asians are good friends but not boyfriends'. Similarly, Rick, a well educated participant with a highly professional job who moved to Canada about a year ago, did not consider himself culturally inferior, arguing that unlike white guys, Asian men were loyal and thus good for long-term relationships: 'Asian men are slim and cute but not good in bed ... they are lovely. They are better for long-term relationships. They are caring and considerate but they are sensitive too .... sometimes a little sentimental'.

For the participants, physically, Asian men might be seen as unattractive and thus undesirable partners. But by asserting that Asian men have the best personalities and are morally superior to white men (see also Espiritu, 2001) they reframed themselves as desirable (long-term) partners to white men whom both Jack and Rick fancy: 'sure, white men are good in bed but we are loyal, loving, caring, good for long-term relationships and wouldn't ever cheat on you!' As a result, they restored a sense of positive self even though such talks may further reinforce the negative image of gay Asian men as being docile.

The participants also did not see themselves conforming to the stereotypical images of gay Asian men. Referring to himself as someone with a unique style, Jack claimed that he did not posses any 'typical Asian' characteristics, to which rice queens[2] were attracted:

> [Rice queens expect you] to be more Asian, thin and short ... and everything I am not ... then there are Asian people who try really hard to be who they aren't ... like really whitewashed ... I dunno what category I am in. I have my own style. I am outcast amongst an outcast.

Like Jack, Tom preferred his stocky build, which he thought made him different from other Asian men: 'I don't really fit the typical Asian description... smaller, submissive and smooth. I am taller, 5 ft11 ins, bigger, 180lbs. I prefer it'.

For them, they were not only physically, but also behaviorally different from the 'typical' gay Asian men. Brian proudly claimed that he was, by being at the opposite end of the spectrum, setting a good example to (passive and submissive) gay Asian men: 'I am probably the other spectrum of the Asian stereotypes though . . . the militia, ever all about the honor, serious yet calm and calculating Asian . . . though I am a good guy. They need a superior macho icon like me'. Eric, a young Canadian-born participant in his early 20s, admitted that he liked white guys but insisted that he did not behave like other Asian men who were so desperate to have a white boyfriend and did not care if he was old, fat and ugly: 'Maybe to get the diehard bananas to chill a little . . . not EVERY Asian is going after whites like they are!!' Eric, unlike older gay Asian men and those gay Asian men who spoke little English or who were unfamiliar with the Canadian cultures, seemingly had more options when it came to dating and, as he claimed, did not need to settle with a rice queen – a condition that mediated Eric's experience and articulation as being gay and Asian.

Of course, the participants can never completely or freely redefine themselves, as their (Asian) body has already been defined socially, and their articulation is always constrained by the ways in which their body is constituted historically, culturally and discursively (see Butler, 1990). But by reframing their bodies, which are culturally stigmatized, they make themselves somewhat 'anew', someone who not only possesses 'good' Asian qualities but also, and perhaps more importantly, does not act or look like the 'typical' gay Asian men whom they consider to signify femininity, passivity and submissiveness. They thus create a 'new' sense of self, representing themselves as different but 'good' gay Asian men. This positioning, one must not assume, is either fixed or permanent; it is relational and constantly changing – a process in which they continuously (re)position themselves against social stigma while negotiating their subjectivities along with the negative feelings (resulting from the stereotypes) as well as the positive feelings about themselves being gay Asians.[3]

However, by redefining themselves positively, they also simultaneously, though unwittingly, (re)produce the 'Other', the 'bad' gay Asian men whom they blame for maintaining 'the negative stereotypes about gay Asian men. For example, Brian, a young participant in his early 20s who called himself 'a superior macho icon', remarked, 'I just find that . . . my Asian counterparts tread on some stereotype without thought to what they do to the collective'. Similarly, Ted, a participant who was highly professional and economically privileged, charged the sex industry in Thailand (an economically disadvantaged country) for upholding the stigma: 'I hate the sex industry in Thailand making a bad name out of Asians'.

Negotiating social stigma associated with their racial preference

When it comes to dating, physical attractiveness is frequently the most important criteria for gay men (Sergios and Cody, 1985). Unsurprisingly, in a culture that idealizes tall, masculine, young, white, male, able bodies while pathologizing other types of bodies as 'unattractive' and 'abnormal', these physical traits are commonly used as a measure of attractiveness. As Sam stated, 'If it's for sex, then I have to be attracted to them but for friendship I don't have any limits . . . younger than me, white, fit . . . looks play a big factor in that [sex and dating]'.

In fact, according to our participants, this racial preference is common among gay Asian men. Robert, for example, found 'that a lot of Asian guys on-line just want whites'. Rick spoke about his difficulty of finding a white boyfriend, which he thought was mainly because there were many more Asian men seeking white men as their partners than vice-versa and only a few young white guys interested in Asian men: 'There is definitely a lot of trouble [finding a white guy] who likes Asians, only a few people and mostly old guys' (see also Choi et al., 1995; Matteson, 1997; McCaskell, 1998; Poon, 2000; Ridge et al., 1999).

However, such desire is highly stigmatized both in the mainstream gay community and in the gay Asian community. As Jackson writes:

> Dominant discourses within White gay cultures stereotype Caucasian-Asian homoerotic relations in terms of two opposed models: firstly, the exclusion of Asian men as erotically attractive and, secondly, the fetishisation of Asian men as the only possible objects of erotic interest. (2000: 183)

Similarly, in 'Power and Intimacy: Caucasian/Asian Gay Relationship as an Indicator of Self-Oppression among Gay Asian Males', Philip Tsui writes:

> In an effort to salvage their self-esteem and to maintain their ego integrity, it appears that some gay Asian men adopt a 'colonial-slave mentality' with its prominent feature being the identification with the oppressor. They learn to accept the categorization of inferiority, and to externalize their self-denial and self-hate (through dissociating themselves from other gay Asians and rejecting them as possible partners for intimacy). (1986: 69)

In contrast, Asian/Asian relationships are seen by some gay Asians as a challenge to the racial hierarchy (particularly of beauty) and a form of collective empowerment (see Chang, 2001; Kapac, 1998).

Interestingly, however, almost all of our participants who dated only white men did not consider their erotic desire for white men as a sign of self-hatred. Nor did they think that dating an Asian man is a sign of self-love or a symbol of self-acceptance. Instead, they saw their erotic desire for a white man as arising from a personal preference: there is no politics, no self-hate; it is simply an attraction. Victor, for example, proudly admitted to his liking for white men and simply explained it as a personal choice: 'hmm . . . hard to explain. It is just an attraction'. Even Jack, who admitted that the negative stereotypes about gay Asian men and his experience of rejection by white men had affected his self-esteem greatly, insisted that his preference for white men was not due to his self-hatred, but because of his attraction to the opposite: 'I like different and if you're attracted to the opposite, why would you be attracted to someone who is Asian like yourself'.

However, this talk of opposite attraction refers to whiteness, not blackness. For example, Justin found that brown hair and blue eyes turned him on but black hair and brown eyes did not: 'I admit I have leaning towards white guys . . . Irish especially . . . something about brown hair blue eyes turns me on'. Like Justin, Tom was attracted to men who were his 'height or taller, average to stocky in size', with 'facial hair', 'around 30 to 35 years old' and 'usually Caucasian'. 'Asians do not usually fit the bill', he said.

What they found problematic was not their erotic desire for white men, but rice queens' fetishization of Asian bodies that are stereotyped as skinny, hairless and culturally feminized. They saw rice queens as 'evils' that not only fetishize the Asian body, but also prey on young Asian men. For example, Jack, a 19-year-old participant, considered rice queens old, unattractive and obsessive with the Asian body, which he found disturbing:

> Then there are rice queens. From my experience, really, really old men prefer Asian boys . . . I don't think rice queens find me that great either cuz I have this feeling they're into the stereotypical look . . . cuz they only like Asians. Most rice queens are like totally hooked into the Asian culture and it kinda disturbs me.

Like Jack, Robert, a participant in his late teens, thought of rice queens unfavorably, using terms such as 'predators' and 'fetishists' to describe them: 'I think [rice queens] mostly are predators, and they only see you as an object. I think in their mind, it is a matter of conquest, like they don't even care about who you are as a person'.

For the participants, rice queens are typically 'middle-aged or older, wealthy, and overweight' who, writes Daniel Gawthrop, are 'drawn to youthful, androgynous features typical of the "Oriental" look: smooth brown skin, black hair, and broad faces with high cheekbones, elongated ("slanted") eyes, and porcelain-perfect lips'. Their lovers are 'sleek, feminine, servile, and passive in bed' – a preference, Gawthrop goes on to write, 'based on inequality' and 'ethnic fetishism':

> For some white men, the appeal is transgressive: Asian guys are a turn-on because their boyish looks, regardless of their actual age, allow for paedophilic fantasies that can be acted upon with exhilarating results – but without breaking the law. For others, the appeal is rooted in culturally determined, essentialist notions of Asian passivity or femininity. Asian guys are seen as more 'gentle' or agreeable than white guys, so an interracial match is seen as complimentary. (Gawthrop, 2005)

Unsurprisingly, those 'saddled with the label are often charged with neo-colonial racism' (Gawthrop, 2005: 9–10). For these reasons, few want to identify themselves or to be identified as rice queens or their lovers; 'everyone wants to show that he's not like the rest of them' (McCaskell, 1998: 45).

What are deemed equally bad, if not worse, are those Asian men who go out with rice queens. These Asian men are seen as feeding on the stereotypes. For many of the participants, they are not only desperate, but also 'drop dead pathetic', making a bad name for 'the Asian'. For example, Nick, a 26-year-old participant who saw himself as masculine, commented, 'There's a lot of Asians who just go for whites and I don't think they care if they're older. They only have to be white . . . Asian men are expected to be submissive. I think some Asians set themselves up for that'. Similarly, Sam (22 years old) stated while self-proclaiming himself as a 'top', 'honestly, femme boys turn me off'. He went on to criticize Asian men who date rice queens:

> Many Asian boys are making their life hard. Some are drop dead pathetic. This is all about this 'potato queen' [a term that refers to Asian men who are primarily attracted to white men] issue. Some Asian boys will go after everything that is white meat. Walk the Church Street [the central area of the gay community in Toronto] and you will see . . . I just don't get those Asian boys. I am sure you know what I am talking about. It's all right to have a type but some Asian boys have gone out of control.

By differentiating themselves from those who date rice queens, they consider their erotic desire for white men fundamentally different: Asian men who date rice queens signify a form of sexual colonization and self-hatred but they don't. It is clear for them that dating white men is fine as long as they are not rice queens – in spite of that they may end up dating someone who is characterized as a 'rice queen' by their peers.[4] As Tom, who proudly admitted his preference for white men, said, 'I don't want to be classified or fit into a category. I think that's why I tend to attract guys who are not just into Asians and I prefer that'. He went on to argue that, since he did not look like a 'typical' Asian man, white men who are attracted to him are not rice queens. The line is clear, though unspoken: if you date rice queens, you are 'bad'; 'good' Asian men don't.

We must not take these accounts as indications of their denial or attempts to cover their inner self-hatred. Nor must we simply dismiss them as being afflicted with internalized racism. As Meg Barker states, 'contradictory discourses are interesting aspects inherent in most speech. We all use different rhetorical devices at different times, when we are trying to create specific effects or achieve different ends' (2005: 80). In fact, this discourse enables them to articulate their erotic desire for white men without pathologizing it, turning it from a language that charges with internalized racism and self-hate to one that celebrates difference and diversity. This language is always portrayed as something positive in a society like Canada that embraces the notion of multiculturalism. Erotic desire for white men is reframed in a new, but positive light – diversity, opposite attraction rather than self-oppression. This reframing, however, it is important to note, must be seen as situational and constantly shifting – a process in which they continuously use different discourses to (re)frame and position themselves against the social stigma associated with their racial preference.

For many anti-racist scholars, this kind of multicultural discourse about diversity does not provide any strategy or analysis for structural change; rather, it serves to disguise the underlying structural inequality in our society and thus further perpetuates oppression (see for example James, 1996). We agree with this claim whole-heartedly but we also do not think that there is merely a one-dimensional effect of such discourse; rather, it serves different purposes at different times in different contexts. As Adrienne Chambon and Allan Irving put it: 'Empowering practices can have a down side; they can impose and exclude. Conversely, certain apparently disempowering practices may contain kernels of productivity' (1999: 263). Here, the discourse of celebrating difference and diversity provides a space that enables the participants to (re)articulate and negotiate their racial preference – giving a language to speak about it and reframe the negative connotation associated with it.

Discussion

Indisputably, negative stereotypes have imposed certain meaning onto gay Asians' bodies but they do not simply accept these meanings. As shown in the narratives, they actively resist and, by reframing their culturally stigmatized bodies, regain a sense of positive self. Unlike resistance at a political or structural level, this type of micro resistance – strategies that gay Asian men use to resist at a personal level – is being little discussed and theorized in gay Asian criticism.

The accounts presented here compel us to reconceptualize our theorization of gay Asian experience that commonly centers its analysis on the intersection of 'race' and 'sexuality'. It is clear in the narratives that not only 'race' and 'sexuality', but also other social categories such as class, body characteristics (including age, height and body shape) and education mediate the way in which gay Asian men resist and strategically (re)position themselves against social inequality.

As a result of the changing global economics, many Asian people (unlike before when most of them came to Canada and the USA as indentured workers) are now professionals with a high level of education and great economic power. Together with the discourse of multiculturalism that promotes ethnic-racial diversity, global capitalism has also transformed Asian people from being stigmatized to being celebrated – a commodity, an Asian lifestyle and a model minority (see Chuh, 2003; Nguyen, 2002; Pon, 2000; Poon, 2006). It is these 'new' emerging identities that enable participants to reframe their bodies and rearticulate who they are – such as Ted's charge against the sex industry in Thailand for 'making a bad name out of Asians' and both Jack's and Rick's claim that Asian men are caring, loving, loyal and morally superior to white men.

Likewise, the privileging of youthfulness and tall, masculine bodies in gay cultures has also enabled some Asian men to draw on their physical characteristics and age to reframe and defend themselves against the social stigma associated with their bodies and desire for white men – such as Tom's articulation of his stocky body and Robert's charge of 'old' rice queens. Only by accounting for other social categories can we see how both 'race' and 'sexuality' manifest themselves differently in different contexts and how gay Asian men can occupy multiple, though contradictory, positions at the same time. As Sik Ying Ho and Kat Tat Tsang write in 'Negotiating Anal Intercourse in Interracial Gay Relationships in Hong Kong':

> The categories of 'race' and 'ethnicity' intersect with other social categories such as class, gender, age, education, etc. to create a unique engagement. The articulation of power is contingent upon site-specific combination of these social categories. In some situations, the 'active–passive' sexual roles follow the lines of 'colonizer–colonialized' and in other situations the opposite happens. In many instances, age and wealth and institutional forms of power are more important than race and colonial status. (2000: 318)

Following Frantz Fanon's work, Jeannette Mageo and Bruce Knauft remind us that domination and oppression are not simply physical acts that superimpose themselves onto our physical spaces but also, and perhaps more importantly, define us, imposing certain meanings onto our body that affect us psychologically:

> One of the most crushing abilities of political oppression is to effect psychological forms of alienation in which people lose loyalty to themselves – to their profoundest feelings and to their love of self. (Mageo and Knauft, 2002: 1)

By reframing their bodies, participants reposition themselves against the stereotypes, which enables them to give 'new' meanings to their bodies and see themselves differently. As Pile writes:

> Psychic resistance is not only unconscious . . . [but] also highly dynamic. It can work in many ways to defend people against interventions – whether therapeutic or political – that seek to persuade, to move things on, to enable people to draw new conclusions about the reasons for things. (1997: 25)

Interestingly, however, as revealed in the analysis, such resistance does not necessarily eliminate their negative feelings about other gay Asian men or their own idealization of whiteness but, rather, co-exists with them. In other words, participants do not simply think of themselves either positively or negatively, but both and more: of course, it often depends on the contexts. Self-acceptance thus is not a simple linear process, but a process of ambivalence in which they continuously negotiate and articulate their multiple subjectivities (identities) in the course of everyday life.

Undeniably, by reframing their bodies, participants shift some of their negative perceptions about themselves that result from social stigma and rejections, as shown earlier in both Emmett's and Michael's cases. This strategy of resistance is thus quite liberating and empowering for some. However, as shown in the narratives, it does not subvert or transform but, rather, reinforces what Hall calls 'the contradictions of the binary structure of racial stereotyping' through which gay Asian identities have been constituted historically, culturally and discursively (2001: 342). In other words, by representing themselves positively as a strategy of resistance, participants simply maintain the binary structure of racial stereotyping that (re)produces the 'Other', rather than challenging it.

For the participants, as revealed in the analysis, small, effeminate bodies signify 'abnormality' and 'deviance'. To resist the stereotypes, they impose certain aesthetics onto their bodies – redefining themselves positively. But it is their own anxiety of being considered 'abnormal' that leads to self-regulation that, in turn, (re)produces the 'Other' and drives prejudice. As Ann Stoler writes, using Foucault's text to analyze western imperial racism: 'It is normalization that drives racism' (1995: 135).

Homi Bhabha has long been aware of this problem and, in 'The Other Question', urged us to shift our site of resistance:

> My reading of colonial discourse suggests that the point of intervention should shift from the *identification* of images as positive or negative, to an understanding of the *process of subjectification* made possible (and plausible) through stereotypical discourse. (1990: 71; original emphasis)

Likewise, as Foucault writes: 'Maybe the target [of individual freedom] nowadays is not to discover what we are, but to refuse what we are' (quoted in Rabinow, 1984: 22; see also Foucault, 1988). Hence, to be radically subversive, we need to question – rather than define ourselves as being either good or bad – how the norms (knowledge) subjugate our soul and body or, in Dan Williamson's words, to be 'fully conscious and critically engaged in genealogically rendering how cultures enforce norms and essentialized identities, built around and sustained by the constant and pervasive technocratic bureaucratization of culture as a whole' (2001: 132). In other words, we need to refuse the cultural codes imposed upon our body, and show that we gay Asian men are not just feminine, passive and submissive, but also masculine, active, aggressive, and much more.

Notes

1. Online interviews require participants to have a certain degree of conventional and computer literacy in order to participate. This may have excluded potential participants who have difficulty expressing themselves through writing English and those who are unfamiliar with or have limited access to the internet. Online interviews also tended to be slower than face-to-face interviews; thus, they might have affected the way in which participants responded to our questions. For example, participants who typed slowly often gave short answers. In some cases, this made it difficult to fully explore issues we wanted to discuss in the interview (Mann and Stewart, 2000). Moreover, online interviews limited the way in which participants express themselves and, in some cases, made it difficult for them to describe certain emotions. Furthermore, we could not perceive participants' non-verbal cues (such as facial expressions) and feelings through changes in voice and intonation during the interviews. All these factors might have affected our understanding of the data (Im and Chee, 2001). Lastly, even though we asked participants to ethnically identify themselves and, if available, read their personal profile on Gay.com, the authenticity regarding their 'true' identity could not be verified.
2. The term 'rice queens' refers to men, usually Caucasian, who are sexually attracted to men of East and Southeast Asian origins. However, this term is highly contested and does not apply uniformly to all white men who fancy Asians. For example, a young, tall, well-built, handsome white hunk is often treated quite differently and less likely to be labeled as a rice queen than his older counterparts even though they both like Asian men. Similarly, a white man who is called a 'rice queen' by one person may not be characterized as such by another person.
3. Gay Asians, as Maurice Poon observes, have often 'received unfriendly remarks and experienced hostility when speaking of gay experiences that are not centered around the negative experiences associated with being gay or Asian'. These men are 'labeled as "whitewashed" and dismissed as being afflicted with internalized racism or homophobia' (2006: 42). For example, gay Asians, who are active in the mainstream gay community but disassociate themselves from other gay Asians, are frequently thought of as attempting to 'assimilate' themselves to the 'white' gay community while 'separating'

themselves from their own community, which in turn is argued as an indication of being 'whitewashed'. Those claiming they don't experience oppression are seen as trivializing social violence or being 'in denial', while those who date only white men are considered as accepting the white-as-beauty logic and thus 'internalizing' racism. We found this kind of language not only negative and judgmental, but very disempowering as well. Nor does it help us understand how their experiences, as well as their articulations of being gay and Asian, are mediated through factors such as class, physical characters, age and so on. We are thus very reluctant to use it to describe our participants.

4. Even though the participants claimed that they do not like rice queens, it does not necessarily mean that they would not date a rice queen or someone who is characterized as a rice queen by others. Their talk, it is important to note, does not necessarily reflect their actions or behaviors (see Hoffman, 1999: 679). Moreover, the term 'rice queen', as discussed earlier, is highly contested. Someone who they see as a rice queen may not been considered as such by others. Or, others may not share their perception on someone who they see as not being a rice queen. Similarly, the participants stated that they are not 'feminine' or do not like 'feminine', 'submissive' gay Asians. But, again it does not necessarily mean that their behaviors are not considered 'feminine' or 'submissive', or that they would not hang out with a gay Asian man who is or is perceived as 'feminine' by others. 'Submissiveness' and 'femininity' are all highly contested terms. How 'feminine' is 'feminine'? How 'submissive' is 'submissive'? What these narratives show is the way in which the participants draw on culturally privileging traits (such as youthfulness and masculinity) and different discourses to (re)position themselves against the social stigma and to make their listeners think the best of them. Most people want to be seen as 'normal' and 'good', not as 'crazy', 'weird', 'abnormal' or 'bad'. By reassuring (convincing) others of who they are and who they are not, they are also and, inevitably, reassuring themselves.

References

Adler, Ronald B. and Rodman, George (1991) *Understanding Human Communication* (4th edition). Toronto: Holt, Rinehart and Winston.

Ayres, Tony (1999) 'China Doll – The Experience of Being a Gay Chinese Australian', *Journal of Homosexuality* 36(3/4): 87–97.

Barker, Meg (2005) 'This is My Partner, and This is My . . . Partner's Partner: Constructing a Polyamorous Identity in a Monogamous World', *Journal of Constructivist Psychology* 18(1): 75–88.

Bhabha, Homi K. (1990 [1986]) 'The Other Question: Difference, Discrimination and the Discourse of Colonialism', in Russell Ferguson, Martha Gever, Trinh T. Minh-ha, Cornel West (ed.) *Out There: Marginalization and Contemporary Cultures* (foreword by Marcia Tucker), pp. 71–87. New York: The New Museum of Contemporary Art.

Bloom, M., Wood, K. and Chambon, A. (1991) 'The Six Languages of Social Work', *Social Work* 36(6): 530–4.

Butler, Judith (1990) *Gender Trouble: Feminism and the Subversion of Identity*. New York: Routledge.

Chambon, Adrienne S. and Irving, Allan (1999) 'Conclusion: Issues to Look Forward To', in Adrienne S. Chambon, Allan Irving and Laura Epstein (eds) *Reading Foucault for Social Work*, pp. 259–67. New York: Columbia University Press.

Chang, Jason (2001) 'The Truth about Gay Asian Men', *aMagazine*. URL (accessed 30 April 2003): http://modelminority.com/modules.php?name=News&file=article&sid=112

Cho, Song (ed.) (1998) *Rice: Explorations into Gay Asian Culture + Politics*. Toronto: Queer Press.

Choi, K.H., Coates, T.J., Catania, J.A., Lew, S. and Chow, P. (1995) 'High HIV Risk among Gay Asian and Pacific Islander Men in San Francisco – Letter to the Editor', *AIDS* 9(3): 306–7.

Chuang, Kent (1999) 'Using Chopsticks to Eat Steak', *Journal of Homosexuality* 36(3/4): 29–41.

Chuh, Kandice (2003) *Imagine Otherwise: On Asian Americanist Critique*. Durham, NC: Duke University Press.

Clegg, Stewart (2002 [1989]) 'From Frameworks of Power', in Mark Haugaard (ed.) *Power: A Reader*, pp. 249–73. Manchester: Manchester University Press.

Davis, Kathy (2002 [1991]) 'Critical Sociology and Gender Relations', in Mark Haugaard (ed.) *Power: A Reader*, pp. 209–24. Manchester: Manchester University Press.

Davis, M., Hart, G., Bolding, G., Sherr, L. and Elford, J. (2003) 'Appraising Internet Relay Chat as a Technique for Conducting Qualitative Interviews about HIV Risk', paper presented in the STD/HIV Prevention on the Internet Conference, Washington DC, 25–27 August.

Drummond, Murray J.N. (2005) 'Asian Gay Men's Bodies', *Journal of Men's Studies* 13(3) 291–300.

Eng, David L. and Hom, Alice Y. (eds) (1998) *Q & A: Queer in Asian America*. Philadelphia, PA: Temple University Press.

Espiritu, Yen Le (2001) '"We Don't Sleep around Like White Girls Do": Family, Culture and Gender in Filipina American Lives', *Signs: Journal of Women in Culture and Society*, 26 (2): 415–40.

Eysenbach, Gunther and Wyatt, Jeremy (2002) 'Facilitating Research via the Internet', in Bruce C. McKenzie (ed.) *Internet and Medicine* (3rd edition), pp. 221–5. New York: Oxford University Press.

Foucault, Michel (1988) 'Practicing Criticism', in Lawrence D. Kritzman (ed.) *Michel Foucault: Politics, Philosophy, Culture – Interviews and Other Writings 1977–1984*, pp. 152–6. New York: Routledge.

Fung, Richard (1996 [1991]) 'Looking for My Penis: the Eroticized Asian in Gay Video Porn', in Russell Leong (ed.) *Asian American Sexualities: Dimensions of the Gay and lesbian Experience*, pp. 181–98. New York: Routledge.

Gawthrop, Daniel (2005) *The Rice Queen Diaries: A Memoir*. Vancouver, BC: Arsenal Pulp Press.

Hall, Stuart (2001 [1997]) 'The Spectacle of the "Other"', in Margaret

Wetherell, Stephanie Taylor and Simeon J. Yates (eds) *Discourse Theory and Practice: A Reader*, pp. 324–43. Thousand Oaks, CA: SAGE.

Han, Chong-Suk (2005) 'Gay Asian-American Male Seeks Home', *The Gay and Lesbian Review* 12(5): 35–6.

Haugaard, Mark (2002) 'Giddens: Introduction', in Mark Haugaard (ed.) *Power: A Reader*, pp. 146–50. Manchester: Manchester University Press.

Hirschmann, Nancy, J. (2003) *The Subject of Liberty: Toward a Feminist Theory of Freedom*. Princeton, NJ: Princeton University Press.

Ho, Petula Sik Ying and Tsang, Adolf Kat Tat (2000) 'Negotiating Anal Intercourse in Interracial Gay Relationships in Hong Kong', *Sexualities* 3(3): 299–323.

Hoffman, Diane (1999) 'Turning Power Inside Out: Reflections on Resistance from the (Anthropological) Field', *International Journal of Qualitative Studies in Education* 12(6): 671–87.

Im, Eun-Ok and Chee, Wonshik (2001) 'A Feminist Critique on the Use of the Internet in Nursing Research', *Advances in Nursing Science* 23(4): 67–82.

Isaac, Jeffrey C. (1992) 'Beyond the Three Faces of Power: A Realist Critique', in Thomas E. Wartenberg (ed.) *Rethinking Power*, pp. 32–55. New York: State University of New York Press.

Isajiw, Wsevolod, W. (1999) *Understanding Diversity: Ethnicity and Race in the Canadian Context*. Toronto: Thompson Educational Publishing.

Jackson, Peter A. (2000) '"That's What Rice Queens Study!" White Gay Desire and Representing Asian Homosexualities', *Journal of Australian Studies* 65: 181–8.

James, Carl E. (ed.) (1996) *Perspectives on Racism and the Human Services Sector: A Case for Change*. Toronto: University of Toronto Press.

James, Carl E. (1999) *Seeing Ourselves: Exploring Race, Ethnicity and Culture*. (2nd edition). Toronto: Thompson Educational Publishing.

Kapac, Jack, S. (1998) 'Culture/Community/Race: Chinese Gay Men and the Politics of Identity', *Anthropologica* 40(2): 169–81.

Khayatt, Didi (1994) 'Revealing Moments: The Voice of One Who Lives with Labels', in Carl E. James and Adrienne Shadd (eds) *Talking about Difference: Encounters in Culture, Language and Identity*, pp. 77–89. Toronto: Between the Lines.

Kong, Travis S. K. (2002) 'The Seduction of the Golden Boy: The Body Politics of Hong Kong Gay Men', *Body & Society* 8(1): 29–48.

Leong, Russell (ed.) (1996) *Asian American Sexualities: Dimensions of the Gay and Lesbian Experience*. New York: Routledge.

Mageo, Jeannette Marie and Knauft, Bruce M. (2002) 'Introduction: Theorizing Power and the Self', in Jeannette Marie Mageo (ed.) *Power and the Self*, pp. 1–25. Cambridge: Cambridge University Press.

Manki, Mohammad A. (2003) 'Power, Subjectivity and Strategies of Resistance: The Case of the Acme School', *Tamara: Journal of Critical Postmodern Organization Science* 2(4): 52–75.

Mann, Chris and Stewart, Fiona (2000) *Internet Communication and Qualitative Research: A Handbook for Researching Online*. London: SAGE.

Mao, L., McCormick, J. and Van de Ven, P. (2002) 'Ethnic and Gay Identification: Gay Asian Men Dealing with the Divide', *Culture, Health and Sexuality* 4(4): 419–30.

Matteson, David R. (1997) 'Bisexual and Homosexual Behavior and HIV Risk among Chinese-, Filipino-, and Korean-American Men', *Journal of Sex Research* 24(1): 93–104.

McCaskell, Tim (1998) 'Towards a Sexual Economy of Rice Queenliness: Lust, Power, and Racism', in Song Cho (ed.) *Rice: Explorations into Gay Asian Culture + Politics*, pp. 45–8. Toronto: Queer Press.

Nemoto, T., Operario, D., Soma, T., Bao, D., Vajrabukka, A. and Crisostomo, V. (2003) 'HIV Risk and Prevention among Asian/Pacific Islander Men Who Have Sex With Men: Listen to Our Stories', *AIDS Education and Prevention* 15 (supplement A): 7–20.

Nguyen, Viet Thanh (2002) *Race and Resistance: Literature and Politics in Asian America*. New York: Oxford University Press.

Pile, Steve (1997) 'Introduction: Opposition, Political Identities and Spaces of Resistance', in Steve Pile and Michael Keith (eds) *Geographies of Resistance*, pp. 1–32. New York: Routledge.

Pon, Gordon (2000) 'The Art of War or The Wedding Banquet? Asian Canadians, Masculinity, and Antiracism Education', *Canadian Journal of Education* 25(2): 139–51.

Poon, Maurice Kwong-Lai (2000) 'Inter-Racial Same-Sex Abuse: The Vulnerability of Gay Men of Asian Descent in Relationships with Caucasian Men', *Journal of Gay and Lesbian Social Services* 11(4): 39–67.

Poon, Maurice Kwong-Lai (2006) 'The Discourse of Oppression in Contemporary Gay Asian Diasporal Literature: Liberation or Limitation?' *Sexuality and Culture* 10(3): 29–58.

Poon, Maurice Kwong-Lai and Ho, Peter Trung-Thu (2002) 'A Qualitative Analysis of Cultural and Social Vulnerabilities to HIV Infection among Gay, Lesbian, and Bisexual Asian Youth', *Journal of Gay and Lesbian Social Services* 14(3): 43–78.

Poon, M. K. L., Ho, P. T. T., Wong, J. P. H., Wong, G. and Lee, R. (2005) 'Psychosocial Experiences of East and Southeast Asian Men Who Use Gay Internet Chat Rooms in Toronto: An Implication for HIV/AIDS Prevention', *Ethnicity and Health* 10(2): 145–67.

Rabinow, Paul (1984) 'Introduction', in Paul Rabinow (ed.) *The Foucault Reader*, pp. 3–29. New York: Pantheon Books.

Ridge, D., Hee, A. and Minichiello, V. (1999) '"Asian" Men on the Scene: Challenges to "Gay Communities"', *Journal of Homosexuality* 36(3/4): 43–68.

Scott, John (2001) *Power*. Cambridge: Polity.

Sergios, Paul A. and Cody, James (1985) 'Physical Attractiveness and Social Assertiveness Skills in Male Homosexual Dating Behavior and Partner Selection', *Journal of Social Psychology* 125(4): 505–14.

Simons, Jon (1995) *Foucault and the Political*. New York: Routledge.

Strauss, Anselm and Corbin, Juliet M. (1998) *Basics of Qualitative Research: Techniques and Procedures for Developing Grounded Theory* (2nd edition). Thousand Oaks, CA: SAGE.

Stoler, Ann Laura (1995) *Race and the Education of Desire: Foucault's History of Sexuality and the Colonial Order of Things*. Durham, NC: Duke University Press.

Sweely, Tracy L. (1999) 'Introduction', in Tracy L. Sweely (ed.) *Manifesting Power: Gender and the Interpretation of Power in Archaeology*, pp. 1–14. New York: Routledge.

Tsui, Philip (1986) 'Power and Intimacy: Caucasian/Asian Gay Relationship as an Indicator of Self-Oppression among Gay Asian Males', *Journal of the Asian American Psychological Association*: 59–61.

Westwood, Sallie (2002) *Power and the Social*. New York: Routledge.

Williamson, Dan (2001) 'Resistance, Self-Fashioning, and Gay Identity', in Steve Martinot with Joy James (ed.) *The Problem of Resistance: Studies in Alternate Political Cultures*, pp. 119–34. New York: Humanity Books.

Witter, Caroline Posadas (2003) 'Stereotypes Influence Prejudices towards African-Americans and Mexican-Americans', paper presented at the 2003 Southern Sociological Society Conference, Atlanta, Georgia, March 26–30.

Young, Iris Marion (1992 [1988]) 'Five Faces of Oppression', in Thomas E. Wartenberg (ed.) *Rethinking Power*, pp. 174–95. New York: State University of New York Press.

Biographical Notes

Maurice Kwong-Lai Poon is Lecturer in the School of Social Work at York University and Chair of the Advisory Research Committee at Asian Community AIDS Services. His current research focuses on gay Asian diasporal identities and politics, the social construction of violence in gay intimate relationships, and HIV/AIDS issues related to Asian men who have sex with men (MSM). He is active in the development of human services for the gay, lesbian, bisexual and transgendered Asian community in Toronto. *Address*: School of Social Work, Kinsmen Building, York University, 4700 Keele St., Toronto, Ontario, Canada M3J 1P3. [email: mk/poon@yorku.ca]

Peter Trung-Thu Ho, MSW, is an Education and Outreach Worker at Asian Community AIDS Services. Together with Maurice Poon and other associates, he has developed many HIV prevention strategies for the Asian MSM population in Toronto, Canada. *Address*: Asian Community AIDS Services, 33 Isabella St., Suite 107, Toronto, Ontario, Canada M4Y 2P7. [email: ho2@sympatico.ca]

CHAPTER 13

MANAGING A DEVIANT IDENTITY

Modes of Suburban Gay Identity

WAYNE BREKHUS

Absence of everyday social contact with minorities often results in negative stereotypes. Derisive terms take the place of personal knowledge of such people. Typifications of unconventionality become amplified and oversimplified conceptions, often caricatures. But, like everyone else, such individuals have to adapt to their social situation. And where the conventional social world sees but a singular type of person, studies reveal variety in both their adaptations and in their commitment to the unconventional identity.

Where the straight world lumps all gays in one category, Wayne Brekhus shows that one simple categorization does not fit them all. He finds variety, not uniformity, among gays. Variations in the frequency, intensity, and duration of contact with gay and straight worlds produce different ways to manage a gay identity. Brekhus finds three modes of identity management among suburban gays. The first is the "full-time gay." This type of suburban gay identifies with and participates the most in the highly visible gay world. Gayness is both his key status and his core self. The "integrated" gay participates more fully in the straight world. His gayness is invisible except to a few intimates and it shares key status with nonsexual aspects of his life. The gay "commuter" makes sporadic visits to gay locations in the city, displays his gayness there, only to return to the straight world where he passes as a straight person. Each of these types of suburban gay, ethnocentrically, claims superiority for their way of managing a gay identity.

The three identity management types are those of (1) lifestylers, (2) commuters, and (3) integrators. Underlying each of these identity types are different configurations of *identity duration* (the degree to which identity is distributed across various times and spaces of one's life), *identity density* (the degree to which identity is packaged and presented in a concentrated or diluted form), and *identity dominance* (the product of duration times density, or the degree to which an identity attribute occupies one's whole self).

The first type is the *identity lifestyler*. Gay lifestylers live openly in gay-specific ghettos and organize their life around their marked status. They keep their markedness on "high volume" and do it virtually all the time. They have a high-density, high-duration gay identity. Metaphorically, they are 100 percent gay, 100 percent of the time. Lifestylers take on the grammatical centrality of gayness as a *noun*.

The second type is the *identity commuter*. Gay commuters treat their gay identity as a *verb*. They live other parts of themselves in heterosexual

Reprinted from *Peacocks, Chameleons, Centaurs: Gay Suburbia and the Grammar of Social Identity,* pp. 28–29 and 98–107, by permission of the University of Chicago Press and the author. Copyright © 2003 by the University of Chicago Press.

space and travel to identity-specific spaces to be their "gay self." They are weekend warriors and nighttime commuters to the gay community. In much the same way that work commuters travel from the suburbs to the city to do work on weekdays, identity commuters travel from the suburbs to the city to "do identity" on weekends. For them, gayness is a *temporary master status* that they turn on and off depending on their social environment—in gay spaces they often turn their marked identity to high volume, but outside these spaces they turn it off completely. They commute to ghettos or bars and play up their markedness in marked social spaces. They submerge markedness and foreground their unmarked characteristics everywhere else. They limit open expression of their gay identity to very few spheres of their lives. Thus, it is a low-duration, high-density gay identity. Metaphorically, they may be something like 100 percent gay, 15 percent of the time. (These percentages are illustrative and not literal. I use 15 percent as the illustrative number for duration because it conveys a small but still significant portion of one's time. This also represents the kind of duration to an identity that is consistent with a weekend identity commitment. Time commitments as low as 1 percent or as high as 40 percent can also be seen as within the duration range of a real-life commuter.)

The third type is the *identity integrator*. The gay identity of gay integrators is an *adjective*. They live openly in heterosexual space and integrate their gay identity into living in a heterosexualized world. In fact, many integrators do not often travel to gay-specific spaces. While they are not closeted, they also do not play up their markedness as lifestylers or commuters do. Unlike commuters, their gay identity is turned on all of the time, but it is at a low volume. Their marked trait becomes just one of a number of facets by which they organize their life and identify themselves. Their gay identity is not entirely off, nor is it on high density. They do markedness on low volume and spread it across all spheres of their lives. They present a low-density, high-duration gay identity. Metaphorically, they might be 15 percent gay, 100 percent of the time. (Again both percentages are illustrative rather than literal. The 15 percent figure for density conveys an attribute that defines only a fraction of the self but a fraction that is still significant.) . . .

IDENTITY NATIVES AND IDENTITY TOURISTS: DURATION DISPUTES OVER GAY IDENTITY

Lifestylers believe that one's commitment to an identity should be undivided. They possess a greedy identity—an identity that requires exclusive and undivided identity work. Central to the lifestyler view is an identity politics model that considers gay invisibility a defining factor in the oppression of gays. From this perspective, high-duration gay visibility is critical to overall cultural visibility and acceptance. To the lifestyler, the commuter who resists high-duration gayness is part of the problem, for his invisibility perpetuates the oppression of gays. . . .

The lifestyler perceives those who do an identity only part-time as uncommitted opportunists who merely "play" the identity when it is safe to do so. The lifestyler views himself as an identity native and the commuter as a mere identity tourist. This divide can also be seen between lifestylers and commuters in the lesbian community. One of Stein's (1997:162) informants, for instance, accused some women of lesbian "tourism" or ideological "play," asserting that they were fakes who were merely posing as lesbians on low duration. From the lifestyler's corner, the commuter is a poseur—a weekend warrior to a gay identity who only "plays the identity" because he lacks the full commitment to "live it." The lifestyler dismisses commuters as "pretenders" or "wannabes," or worse yet as "hypocrites" because they show their "true colors" only when it is safe to do so.

Some gays in New York City, for instance, refer to suburban identity commuters as *tunnel and bridge gays*. From the New York lifestyler's perspective the tunnel and bridge gay is tainted by his association with straight suburbia and his lack of sophistication regarding the culture among natives.

From the perspective of some urban gays tunnel and bridge gays, like "tourists" to any culture, do a poor job of following the local customs and thus fitting into the native culture. Indeed, those very customs may be there for the express purpose of separating the inner circle from the wanna-bes.

Part of the urban lifestyler's objection to tunnel and bridge gays is that they turn off their gay identity as soon as they pass back through the tunnel into the heterosexual world. The commuter's gay life consists of fleeting weekend bursts of gay identity in the city followed by long stretches of mundane suburban living. The duration of his gayness is too short for the lifestyler to take him seriously. From the lifestyler's perspective he won't be a "true gay" until he "goes native" and abandons the straight world. As long as he is only commuting to or vacationing in the city he is not a true insider but merely a curious spectator to the gay subculture. The lifestyler expects total immersion in gay culture not just on the weekends but all of the time. Moreover, the lifestyler views anyone who is only a part-time gay as uncommitted to his "core self." The lifestyler views individuals who repress this all-important part of their identity at some times and in some places as "self-hating." Signorile . . . , for instance, refers to the closeted gay, in which category his definition includes the commuter, as a "self-loathing homosexual" who allows his entire self to be dominated by oppression most of the time. The gay lifestyler, who views gayness as a noun, sees any act that does not openly proclaim one's gayness as a betrayal of one's true inner "core self." He believes the chameleon has sold his soul and betrayed his true colors every time he turns in his bright urban gay colors for the bland straight colors of mundane suburbia. The lifestyler views the identities of commuters as impure and inevitably tainted by their high-duration contact with the profane non-gay world. For instance, Levine's . . . informants perceived suburban commuters, who failed to adopt the full dress code of clone culture, as visual pollutants to their "authentically gay" environment. Some queer theorists have adopted the lifestyler's position on duration as their own analytic position, suggesting that the closeted homosexual, again defined to include commuters, suffers from a polluted self-image.

The commuter views his duration differently. Rather than seeing low duration as a lack of authenticity or commitment, he sees it as an abundance of mobility. From the commuter's view, "full-time gays" (lifestylers and integrators alike) lack flexibility. For instance, identity commuters express the view that there is a difference between a "part-time queen" and a "full-time queen." As Bill states:

Now there's a difference between [being] a queen and [acting] queeny. A person can be queeny but not necessarily a queen. I mean a queen is a queen all the time, and that's what they are. They're defined by it. Those people who are Joe Blow but they can be queeny at times, that's a little different. They can act effeminate at times but a queen has made it their life.

For Bill, someone who can be queeny at times is superior to someone who can only be a queen, because the queeny person also has the flexibility to be Joe Blow if he wants to. Although Bill's disdain for "queens" may not be clear from the above quote alone, his attitude toward queens was highly unfavorable. By contrast he had a higher regard for the queeny person because, in his view, the queeny person is engaging in play and thus his "core" is still masculine. The commuter views being defined by a full-time queen status as an unfortunate state of affairs that only afflicts those whose femininity is so overpowering she can never be anything else; her flavor is always the same. In the commuter's eyes the queen is frozen because she can never change into Joe Blow or anyone else; she is one-dimensional, a noun. By contrast the queeny person has the flexibility to slip out of his queeny persona and into other personas, including generic ones. In fact, for some commuters the generic ordinary Joe rather than the gay queen is their default identity. One suburban gay who defined himself as a regular guy, for instance, argued strongly for the low duration of his gayness in saying that "*very few minutes of the day* are defined by who I like to have sex with." Far more minutes a day, he went on to suggest, are defined by his position as an "ordinary guy in an ordinary suburban neighborhood."

The commuter views the full-time gay's sin as his inability to turn off his gayness in nongay spaces. For the commuter the lifestyler's inability to be anything but gay can be embarrassing. As Scott explains:

> *Some people are more tolerant [of queens], and I'm probably a little less tolerant. And it might depend on location. In some gay groups I'm not as put off by effeminate gay men as when I'm in a public restaurant and someone comes up to me who was overly effeminate. [In that case] I might feel a little embarrassed.*

Scott's primary concern is with queen displays that extend into unmarked times and spaces. He tolerates effeminacy in gay spaces "where they belong" but is embarrassed when such displays cross over into the nongay realm. For him and other commuters, the always effeminate gay man threatens his boundaries between gay and nongay time and space. Gay commuters worry that effeminate men may hinder their own mobility to move between a gay self in marked times and spaces and a nongay one outside those spaces.

Brian worries about the potential staining effect of spending too much of his time around gay lifestylers. A new traveler to the gay scene, Brian feels he can tell some people are gay by that little flamboyancy in their voice that he believes they have acquired from hanging around too much in all-gay environments. He likens picking up a "gay accent" from hanging around in New York's gay spaces to picking up a regional accent:

> *I don't want to say that people choose to talk feminine, but when it comes right down to it, it's learned to talk that way. Kind of like if you were a Texan in New Jersey you would lose your accent and if a New Jerseyan went to Texas they would gain an accent. So I think it's kind of like gaining an accent. So that's why I'm not particularly fond of hanging around very feminine people [too long]. I mean they're fun and everything but—yeah, I talk a little bit more feminine joking around with some of my [gay] friends at times, but then I'm like, "stop that."*

Brian is content to be a traveler to peacock gay spaces and to even play it up a little himself when he is there, but he fears that too much time in gay space will cause him to pick up the "native accent" of gay lifestylers. Since he still has to turn it off in his life outside gay space, he worries about the potential spillover effect of spending too much time in the presence of queens and effeminate gay men. Like other commuters, he wants to be able to move between the two identities and their separate spaces.

The commuter views full-time gayness as a problem because it invades inappropriate times and places. For the commuter, weekends, nighttime, and vacations are the proper times to let gayness take over and become a master status. The commuter believes other times should be reserved for his unmarked identities. Commuters I spoke with, for instance, often disparaged gay politics as something for people "whose only goal in life is to live a gay lifestyle." They argued that they themselves held "real jobs" and thus were above what they saw as the immaturity of full-time gay activists. Andrew says, for instance, that while he agrees with the goals of gay activists, he thinks it is "silly how much *time* they let it consume." He goes on to say that people who take their jobs seriously don't have time to be "politically gay all of the time":

> *People who are working just don't have time to be on these panels or [in] these groups fighting for causes.... Gay politics never really concerned me. It never really interested me, and I never felt the desire to do anything about it. It's nice for these people who are graduate students and they don't really know what to do with their lives. I know a lot of people like this. This is their goal in life: to be gay, to be politically gay. But, ah, my life goes on.*

For Andrew and other commuters, the problem with gay activists is that they devote too much duration to their gay identity and thus do not leave enough duration to other identities such as their occupational identities. For many gay bar patrons like Andrew "being gay" is a nighttime activity. Any activist gay politics would likely extend beyond the duration and the confines of the bar scene and thus impinge on activities reserved for his diurnal self. Underlying Andrew's distinction between "people who are graduate students" (student lifestylers with the auxiliary characteristic of political activism) and "people who are working" is a temporal distinction between the full-time gay and the part-time gay. This same concern among commuters that a full-time gay identity pushes out other important identities came through when Bill explained his view that a queen is superficial because, unlike a queeny person, she can never talk about other facets of self: "queens are superficial because the only thing they ever talk about is gay life; they never talk about anything of substance like work or sports." Bill equates work and sports with substance because they are ingredients beyond one's nighttime self and beyond the auxiliary characteristics of gayness.

Like the lifestyler, the integrator has a problem with the part-time aspect of the gay commuter's identity. However, while the lifestyler complains that the gay commuter is only being true to himself in gay space, the integrator complains that the commuter is never being true to himself because he is always bracketing a part of himself. For the integrator neither suppressing one's gay identity nor foregrounding it and suppressing one's diurnal self are true displays of oneself. Warren expresses the integrator view:

> *I don't like that some gay people I know, they are in the closet and they hide it . . . and they'll even say homophobic stuff just to fit in . . . and then—but then you go to the bar and they are the biggest queen in the bar. I mean if you're going to be gay why not just be gay? I mean just be yourself and be gay. Don't hide it, but don't make a big deal of it either. If you don't act [the way you act in a gay bar] normally, what makes you think you should act that way just because you are in a gay bar?*

For Warren and other integrators, the commuter is problematic because he overcompensates for low duration with high density. He questions why the commuter cannot commit to one "core self" instead of shifting between two identity poles. He wonders why the commuter cannot just be one everyday gay self all of the time instead of a part-time everyday self and a part-time gay self. In the integrator's view, the commuter is untrue because his whole self is never present. He is always bracketing something off and hiding it from view. In contrast, all of the integrator's identity attributes are always accessible. Whereas the commuter divides his different identity commitments into separate temporal and spatial domains, the integrator mixes all of his commitments together into every minute of his life. One might think of the commuter's identity plate as resembling a TV dinner (all separate dishes compartmentalized into separate spheres so that only one dish can be accessed at a time), while the integrator's plate resembles a salad or a soup, where everything is mixed together. (The lifestyler's plate would have one giant habanero pepper—all spice with no moderating flavors.)

An integrator's view on the durational problem of commuters is exemplified by Bawer's complaint against gays whom he perceives as choosing to "shout on high volume" one day a year rather than put in the time to convince people "quietly" 365 days a year—that is for a display of too short duration and too high density:

> *At Gay Pride Day Marches, some gay men and lesbians, like the Stonewall Rioters, have exposed America to images of raw sexuality—images that variously amuse, titillate, shock, and offend while revealing nothing important about who most of those people really are. Why, then, do some people do such things? Perhaps because they've been conditioned to think that on that gay high holy day, the definitively gay thing to do is to be as defiant as those heroes twenty-five years ago. Perhaps they do it because they can more easily grasp the concept of enjoying one day per year of delicious anarchy than of devoting 365 days per year to a somewhat more disciplined and strategically sensible demonstration designed to advance the causes of respect, dignity, and equality. . . .*

Interestingly, both lifestylers and integrators attack commuters for their short binges of gay display. They differ, however, in that the lifestyler views the bursts of gayness as the commuter's true colors and the rest of his duration as problematic, while the integrator views the commuter's short binges as the problem and his "364-day self" outside the parade as closer to his true colors.

On the issue of duration, commuters favor a mobile self while lifestylers and integrators favor a unitary self. For the commuter the self is flexible and fluid enough to blend in anywhere. The lifestyler conceives of the unitary self as organized hierarchically around a master status with all other facets of self as auxiliary characteristics. The integrator's unitary self, by contrast, is composed of constant adjectives arrayed in a flat, modular non-hierarchical way around the nebulous label of "human being." The lifestyler and the integrator cannot understand the commuter's constantly changing self since it shows little commitment to a "core self" of any kind. The commuter, by contrast, cannot understand the frozenness and inflexibility of his lifestyler and integrator counterparts, who refuse to adapt and respond appropriately to changes in their social environment.

# Being Gay and Jewish: Negotiating Intersecting Identities

Randal F. Schnoor*
York University

Due to the emphasis on "traditional" gender roles, the "nuclear family," procreation and conservative religious values, many gay and lesbian Jews feel a sense of alienation from the Jewish community and develop an ambivalent or conflicted relationship about their own Jewish identity. As a result, gay Jews often struggle to find ways to successfully negotiate their ethno-religious and sexual identities. Based upon in-depth interviews of thirty gay Jewish men in Toronto, this work offers a case study to empirically and theoretically explore the varied experiences of these intersecting identities for this under-studied population. Recent research on other ethnic minority gays and lesbians tend to simplify this question by suggesting that the minority gay individual will simply choose to prioritize one of these identities while repressing the other. Building upon studies of gay Christians that stress more fluid, dynamic and evolving approaches to identity construction, this paper underscores the complexity and variability of this phenomenon as it applies to gay Jews.

I'm tired of fighting. I'm exhausted. I spend most of Shabbat sleeping because I'm just exhausted with the daily fight of my identity. It's hard enough having one minority status to negotiate in a society, but to have two, it's exhausting.
Traditional gay Jew in his forties

INTRODUCTION

Due to the emphasis on "traditional" gender roles, the "nuclear family," procreation and conservative religious values, many gay and lesbian Jews feel a sense of alienation from the Jewish community and develop an ambivalent or conflicted relationship about their own Jewish identity (Fishman 2000:106-109; Schnoor 2003). In addition to the difficulty of homophobia in the Jewish community, gay or lesbian Jews cannot presume full acceptance from the broader queer community, as movements that focus on sustaining a collective identity for an oppressed

---

Direct correspondence to: Randal F. Schnoor, York University, Toronto, Ontario. Email: randal.schnoor@primus.ca. I wish to thank Nancy Nason-Clark, the editor of Sociology of Religion, as well as two anonymous reviewers for their helpful comments. This research was supported by a grant from the Memorial Foundation for Jewish Culture. Additional support came from McGill University through the Max Bell Fellowship as well as a doctoral fellowship.

group sometimes overlook the concerns of sub-groups found within the larger movement (Eder, Staggenborg and Sudderth 1995:489; Goffman 1963:138). As a result, gay Jews often struggle to find ways to successfully negotiate their ethno-religious and sexual identities. Based upon in-depth interviews of thirty gay Jewish men in Toronto, this work offers a case study in which to empirically and theoretically explore the varied experiences of these intersecting identities.

It is useful here to review the literature that currently exists on the subject of gay and lesbian Jews. While there is a collection of fiction on the subject (Raphael 1996; Schimel 2000; Tulchinsky 2000), a significant amount of material that offers personal narratives or life-histories (Alpert, Elwell and Idelson 2001; Balka and Rose 1989; Brown 2004; Dworkin 1997; Fink and Press 1999; Moore 1995; Shneer and Aviv 2002; Torton Beck 1989), and a recent body of material that challenges or expounds upon Orthodox Jewish theological interpretations of homosexuality (Alpert 1989, 1997; Greenberg 2004; Levado 1993; Rapoport 2004; Rose 1999), there is precious little in terms of scholarly social-scientific literature devoted to the issue of gay Jews. Shokeid (1995) gives an anthropological account of the gay synagogue in New York City, Cooper (1989) offers a social history of the rise of the gay Jewish movement in terms of its organizational structure, and Walzer (2000) gives a journalistic account of the social and political landscape for gay Jews in Israel. However, with the exception of Mushkat (1999) – a small-scale study based on interview data from a sample of nine Jewish lesbians – and the recent work of wolfman[1] (2002), which focuses on Jewish family issues, there is no empirically based sociological literature that examines micro-issues, such as negotiation of intersecting identities.

As will be discussed later in this work, there are a handful of studies that explore these identity intersections for gays and lesbians of other religious and ethnic groups. This research fills an important void by adding to the sociological literature on identity construction as it pertains to the case of gay Jews.

Interestingly, within the Jewish literature there is considerably more research that concentrates exclusively on Jewish lesbian experiences (Alpert 1997; Alpert et al. 2001; Moore 1995; Mushkat 1999; Torton Beck 1989). There is a distinct gap in the literature regarding gay Jewish male experiences, particularly in the North American context. This work will address this gap by exclusively exploring the issues of men.

Gay Jewish Organizational History in Toronto

As the largest Jewish centre in Canada (180,000 Jews), Toronto has a rich history of organizations which have served gay Jewish interests. As such, it serves as an ideal social laboratory in which to examine the intersections of gay Jewish

---

[1] wolfman does not capitalize his name.

experiences. The first organized gay Jewish entity in Toronto was formed in the mid 1970s and was given the name *Ha'Mishpachah* (The Family) to signify a potential replacement family for those gay Jews who were not receiving sufficient support from their biological families. Certain members of this group had a desire to expand their events to include Friday night religious services. This was the impetus for the formation of the successor group called *B'nai Kehillah* (Children of the Community), a name that more closely resembles those names customarily used for synagogues. The turning point in gay Jewish organizational structure in Toronto came in the early-mid 1980s with the formation *Chutzpah* (Nerve / Gall), a more formalized group that included a constitution, an executive structure and a paying membership. In 1988 the group became an official member of *Keshet Ga'avah* (Rainbow of Pride): *The World Congress of Gay, Lesbian, Bisexual and Transgender Jews*, an international umbrella group of LGBT organizations. At its peak in the early 1990s the group consisted of approximately 150 paying members. Similar to what happened in the late 1970s, the gay Jewish entity of the early 1990s desired to re-name and re-orient itself to a more synagogue-like framework. In 1992, with a new president and executive structure in place, Congregation *Keshet Shalom* (Rainbow of Peace) was created. The group thrived in the mid-1990s, organizing formal prayer services, numerous social events, lectures, Passover Seders and Chanukah parties. Due to lack of Jewish community financial support, the group disbanded in 2001. Meanwhile in March of 2000, the Jewish Campus Services (now called Hillel) of Toronto began a group at the University of Toronto Campus called Jewish Lesbian Gay Bisexual and Transgender Students and Young Professionals of Toronto (JLGBT). This group signals the first time a gay Jewish organization in Toronto is primarily funded through the resources of Toronto's Jewish polity. In 2004, the re-energized campus group changed its name to *Kulanu* (Our Voice).

CONSTRUCTIONIST APPROACHES TO IDENTITY

Building upon the work of Barth (1969), Gans (1979) and others, much current research on ethnic identity in the social sciences rests upon a model of ethnicity as a socially constructed phenomenon. Rather than having a fixed, primordial status, ethnicity is viewed as a construction that is continually negotiated and re-negotiated by the individual (Nagel 1994, 2000; Spector and Kitsuse 1987). Ethnic identifiers can slip in and out of ethnic roles depending on social context (Waters 1990).

Recent work in the sociology of religion has described a similar societal shift in religious identity, a phenomenon referred to as *religious individualism*. Wuthnow (1998:9-10) emphasizes the fact that the individual no longer feels constrained by ascribed characteristics, but rather, as "Sovereign Self," constructs his/her own personal religious identity by pulling together elements from various repertoires. Roof (1999:43) argues similarly that we have become "meaning-making creatures" who selectively choose interpretations to authenticate our own convictions.

Similar constructionist approaches can be found in studies of sexuality. Shaped by the early work of Foucault (1980), a number of recent thinkers within the field of queer theory have challenged assumptions about the nature and content of sexuality. Through the work of Sedgewick (1990), Butler (1990) and others, sociologists have become increasingly sensitized to the conception of fluid and dynamic sexual identities where a strict binary division between the "homosexual" and "heterosexual" no longer holds legitimacy. In his ethnography of suburban gay men, Brekhus (2003), for example, has identified three ideal types of gay identities, where the *gay lifestyler* carves out specialized social enclaves to devote their lives to a celebration of being "all gay, all the time," the *gay commuter* moves in and out of gayness as a temporary status, performing a gay identity only in specific gay environments, and the *gay integrator* combines a gay identity with other identities (ethnic, religious, other) so that no one attribute defines the core self. Working within this framework, "homosexual identity" is de-centred to avoid artificial conceptions of a "master status." Identity is dynamic, capable of shifting and changing depending on context and individual predilections.

STUDIES IN JEWISH IDENTITY

Sociologists of the modern Jewish experience have described a similar de-centring of identity. In the transition from pre-modern, traditional societies to post-Enlightenment, modern societies, Jewish identities have gone through a significant transformation from an identity based on collectivism to an identity based on personalism and voluntarism (Cohen 1999:3-4; Elazar 1999:35). As such, many Jews in the North American context and elsewhere now feel free to appropriate only those aspects of Jewishness that they find personally meaningful (Cohen 1991:27; Cohen and Eisen 2000:7-9; Horowitz 1998:74-75). As part of this meaning-based approach to Jewish identity, is the important realization that a person's experience of Jewishness is a fluid, rather than fixed aspect of their lives. Individuals often go through "Jewish journeys" where Jewish identity evolves over the life course, paralleling growth and personal development (Horowitz 2002:26-27).

While Alba (1990) and Waters (1990) have demonstrated the dilution of loyalties to specific European ethnic identities among whites Americans (Italians, Greeks, etc.), the religious aspect of Jewishness provides an extra dimension, which deepens or enlarges the possibilities of Jewish expression. (Horowitz 2002). Thus, Jewishness is unique among cultural identities in that it can be expressed as both an ethnicity and a religion, sometimes referred to as ethno-religious identity. This variability will be evident among my Jewish respondents and will have a bearing on how these men negotiate the intersections of their gayness and Jewishness.

NEGOTIATING INTERSECTING IDENTITIES

Until the early 1980s the vast majority of the empirical sociological literature on issues surrounding lesbians and gay men in North America described only the "'mainstream white" experience. Since the late 1980s, this gap in the literature has begun to narrow, with a handful of studies on the ethnic minority lesbian and gay experience. Some of these studies address the question of how ethnic minority gay individuals attempt to negotiate this dual minority status. This research often tends to simplify this question by suggesting that the minority gay individual will simply choose to prioritize one of these identities while repressing the other. Morales (1990:228-229), for example, theorizes that the individual will form a primary identification to the ethnic community and develop feelings of anger towards the racism experienced in the gay community. While Greene (1998:48) provides some empirical evidence to support this from her research on the gay black community, evidence of the opposite phenomenon, primary identification with the gay community, is also reported in the literature. We see that Latino gay men (Garcia 1998:109) and Asian gays and lesbians (Chan 1989:19-20), for example, are reported as feeling more comfortable identifying with the gay community than with their ethnic communities.

Recent studies on gay Christians (Mahaffy 1996; Rodriguez and Ouellette 2000; Thumma 1991; Wilcox 2002, 2003) have provided more nuanced examinations of this question by demonstrating a wider range of strategies that gay Christians implement to negotiate their identities[2] as well as the fluid and shifting nature of these strategies. Through the voices of gay Jewish men, this study builds upon these works by underscoring the complexity and variability of negotiating the intersection of gay and Jewish identities.

METHODS

In-depth interviews with thirty gay Jewish men in Toronto served as the source of qualitative data for this study. These interviews were conducted between September 2000 and August 2001. Interview respondents were acquired through snowball sampling, through introductions at gay Jewish events and through referrals from Jewish community networks (friends, family, acquaintances). Gay Jewish respondents were distributed evenly by age, with roughly equal numbers in their teens, twenties, thirties, forties and fifties. One respondent was in his eighties.

---

[2]Rodriguez and Ouellette (2000), for example, identify similar categories to what I discovered among my sample of gay Jewish men. They refer to them as: *rejecting the religious identity*, *rejecting the homosexual identity*, *compartmentalization*, and *identity integration*.

At the time of the interview, respondents demonstrated considerable variability in their expressions of Jewishness, falling into what could be described as two general categories: one third (10) may be classified as traditional Jews (those that place more emphasis on the religious aspect of Jewishness); two thirds (20) may be classified as secular Jews (those that place more emphasis on the ethnic aspect of their Jewishness or place little emphasis on their Jewishness altogether).[3]

Interviews were semi-structured in format. The same groups of central themes were explored with all respondents. Respondents were first asked to describe their Jewish lives and gay lives separately. This was followed by a series of questions that explored the ways that the respondents attempt to negotiate their gay and Jewish identities.

With the exception of four interviews where notes were taken, interviews were recorded on audiotape and later transcribed verbatim. Respondents signed an Informed Consent form before the start of the interview and were assured that the information gathered would be kept confidential and that pseudonyms would be used in the study.

FINDINGS

There are multiple ways that individuals choose to construct their various personal identities. Added complexity exists when considering the intersection of two identities that come into conflict with each other, as can be the case with Jewishness and gayness. Borrowing from, and expanding upon, Brekhus' (2003) model of ideal types of gay identities, I am able to situate my respondents into four primary negotiation strategies. In keeping with the theoretical perspectives which inform this work, it is important to keep in mind that these categories are not meant to be rigid but rather quite fluid and dynamic. As will be shown in the

---

[3]Traditional Jews include both Jews who are strictly religiously observant Orthodox Jews (of which there were three in my sample) as well as those Jews who might not observe all Jewish observances but still maintain "conservative religious values" in terms of placing importance on synagogue attendance, celebrating the Jewish holidays and preserving and perpetuating Jewish religious tradition. Some secular Jews have a strong sense of Jewish identity, but place little importance on these religious concerns, placing greater emphasis instead on matters of ethnic Jewishness (such as Jewish family, food, literature, music, culture, etc.). Other secular Jews may have little attachment to their Jewishness at all. There are, of course, wide ranges of attitudes and behaviors within these categories. These labels should only be understood as approximations. Furthermore, as the paper argues, these identities can evolve and change.

It is helpful to contextualize the Toronto Jewish community. Social scientific research of the Jews of North America has demonstrated that Canadian Jews, in general, are more traditional in their Jewish behaviors than American Jews (Shaffir and Weinfeld 1981:13; Brodbar-Nemzer, Cohen, Reitzes, Shahar and Tobin 1993:43) and Toronto (and Montreal) Jews are more traditional than Jews of other Canadian cities (Davids 1998:212-214).

examples of some of my respondents, these strategies are often temporary or situational, sometimes changing over the lifetime of the individual.[4] One's type of Jewish expression serves as a significant explanatory variable.

1. Jewish Lifestylers

Traditionally minded Jews are the ones more likely to try to repress their gay inclinations so that they do not interfere with their Jewish lives. These are people who consider their Jewishness as one of the key defining features of who they are. Four of my respondents exhibited these qualities of Jewish lifestylers, at least for a certain period of time in their lives. Saul, a traditional Jew, reported:

> If I weren't Jewish, I might be out there fighting for gay rights, but my Jewish identity is far more important to me than my gay identity. And that's something I knew right from [the start]. I did not want to allow my same-gender romantic orientation to affect my passion for Judaism and Jewish life and the Jewish people. I never wanted that to impinge on it, to steal time from it and to even affect it.

While one might guess that being gay would cause a Jew to adopt a more liberal expression of Judaism to accommodate one's homosexuality (a phenomenon that certainly does occur), interestingly two of my respondents consciously chose to become ultra-religious Jewish lifestylers in an attempt to purge themselves of their gay inclination.[5] These respondents also illustrate some shame for being gay. Although somewhat traditionally minded to start with, as a result of the "battle" Nathan was having between his gay and Jewish identities, at a certain period of his life he chose "extreme spirituality because [he] felt guilty for being gay." When I asked him to describe this extreme spirituality he explained that he:

> [became] obsessed about *kashrut* [dietary laws] and *Shabbat* and keeping the holidays. There were a few years [on Yom Kippur] where I made sure I read every word in the *machzor* [prayer book], because it was like the magic potion. If I missed a word it would be like [*gulping sound*] hell to pay.

Surprisingly, even a secular Jew chose the path of extreme religiosity to try to stop himself from being gay. Leonard claims that being gay precipitated his "Jewish journey." He became friendly with some observant Jews at a Jewish summer camp and joined their insular lifestyler community. He claims "It stopped me from being gay for ten years." When I asked him about this seemingly unusual pattern for a gay secular Jew to become extremely observant, he provided the following explanation:

---

[4]The pattern of shifting between strategies is also documented in the case of gay Christians. See Wilcox (2003).

[5]See Mahaffy (1996) and Wilcox (2003) for similar examples of this phenomenon in a Christian context.

> I think it's common for people in adverse situations in their lives to seek out stability. It's the same way old people become religious. Sort of hedging their bets. So I think that religion, maybe twenty years ago more than now, served as a secure footing for people, including me; an opportunity to escape.

I found it interesting to note that both Nathan and Leonard abandoned their Jewish lifestyler identities later in their lives. Nathan's next stage on his journey was to make an extreme switch to a gay lifestyler identity, as will be discussed in the next section. Leonard gradually moderated his strategy to become a gay-Jewish integrator. These two individuals provide a vivid example of the fluidity of the process of identity management. As the literature suggests, identity construction can be temporary or situational. Individuals can go through different stages through their lifespan responding to historical experiences, social relationships and personal events.

2. Gay Lifestylers

Secular Jews are the ones more likely to place strong emphasis on their gayness. Three of my secular Jewish respondents lived in openly gay specific neighborhoods and organized their lives around being gay at "high volumes" and in "high duration." These gay lifestylers did not wish to dilute their gayness by placing importance on other aspects of their social identities. As such, their Jewish identities were significantly de-emphasized.

Though not Jewishly religious in any sense, Russell viewed his Jewishness as a "religion" that could be discarded at any time. His gay identity formed the most essential part of his core self. He explained that:

> To me I'm gay. But that's much more fundamental than being Jewish. I can change my religion but I can't change my sexual orientation.

Internalized distaste for one's Jewishness can play a role in taking on a gay lifestyler identity. Nathan reported that:

> I have never known so many internalized anti-Semites as gay Jews. ... Because of whatever reason – the homophobia in Jewish tradition, misinterpretations of Torah or whatever – it's far easier to just say "fuck it" to the Jewish world and the Jewish identity, and go blazing out of the closet in the gay world. They're not proud of being Jewish.

Nathan himself is an interesting case in point. He managed his identities in extremes. A one-time Jewish lifestyler who became obsessed with Jewish religious observance, in a later period of his life, to his own surprise, he rebelled against Judaism:

> I never thought I'd be the Jew that would say, "Fuck Judaism! It's persecutory." I was always the one who was going to defend the Jewish community at all costs. And here I was actually becoming that person that I thought I'd never be. The person walking out

of the synagogue and saying, "That's a synagogue I'll never go to," questioning a lot about my heritage.

As part of his Jewish rebellion, Nathan became a temporary gay lifestyler, devoting great energy to embracing and celebrating his gayness. Later in his life he adapts yet another management strategy, as will be seen.

3. Gay-Jewish Commuters

For identity commuters the performance of specific identities can be turned on or of depending on the social context. Gay-Jewish commuters travel freely between gay-specific and Jewish-specific settings, but are careful to enact the appropriate identity while submerging the other. By completely compartmentalizing their identities in this way, these commuters are able to enjoy both their gay and Jewish lives while keeping them separate and free of intersection. I encountered four such commuters among my group of gay Jewish men. A clear example of the gay-Jewish commuter can be found in the case of Abe, a religiously observant Jew, who attended prayer services at an Orthodox synagogue on a regular basis but kept his gay identity hidden in this setting. He explained:

> Like many people, I compartmentalize, and being gay is one thing, it's an integral part of me, and I enjoy it to its fullest extent. Being an observant Jew is something that I also enjoy and get satisfaction from, for a number of reasons. The two separately are both things that make me feel good about myself, so I want to participate in them. Bringing the two together gets a bit more problematic. And so that's why you'll see when I'm at the [Orthodox synagogue], I'm not really "out" at all. I want to enjoy the feelings that I get out of the observant experience, the prayer and that communal experience, and so I sort of put the other on hold and enjoy that.

Gary grew up within a very traditional Jewish family in a small Canadian city. Though he desired to please his parents by having a Jewish wedding with a woman, he knew that he only had attractions towards men. Because being gay and Jewish were both very important to him and he could see no way to reconcile his two identities, he felt his best solution in his twenties was to keep the two identities completely separate from each other. He lamented that:

> I didn't know any gay couples. I just knew that to be a gay man was to be ostracized and to be alone and not to be Jewish. For a long time, they were completely separate spheres in my life and I find it hard to reflect on how one can do that, but now I know that people do that all the time.

As will be seen, later in life Gary found ways to feel comfortable about combining his gay and Jewish identities and adopted a gay-Jewish integrator strategy.

4. Gay-Jewish Integrators

Identity integrators view themselves as being made up of a multiple of attributes, where no one social identity assumes a role of "master status" around which his or her life is organized. Gay-Jewish integrators perform both their gay and Jewish identities throughout all their social interactions, but these are expressed at lower intensity level than would be the case of lifestylers. Twenty-four of my respondents exhibited the qualities of gay-Jewish integrators at certain times in their lives. These men brought forward four primary ways of doing this: participation in gay Jewish organizations; challenging Jewish theological perspectives on homosexuality; emphasizing linkages between Judaism and homosexuality; and using "Jewish values" to guide them through the gay world. Again, the different styles of Jewish expression serve as an important explanatory variable that can help to distinguish between different types of strategies. There are both traditional ways and secular ways, as will be shown, to integrate one's gay and Jewish identities.

4a) Participation in Gay Jewish Organizations

Shokeid (1995:239), in his ethnography of the gay synagogue in New York City, described the way that attendance at such a synagogue "was for many an act of restoring their cracked self-image and identity, combining its divided parts into one meaningful identity." My respondents reported similar sentiments. The variety of gay Jewish organizations that have existed since the 1970's in Toronto and elsewhere have provided a focal point for identity integration. My respondents described participation in such organizations as "blending the two nicely," and providing "comfort and sense of belonging."

Nathan, a man who spent part of his adolescence and twenties as both a Jewish lifestlyer and a gay lifestyler, later transferred his passions to a gay-Jewish integrator identity. He became very active in *Keshet Shalom* and *Keshet Ga'avah: The World Congress of Gay, Lesbian, Bisexual and Transgender Jews* attending the latter group's annual conferences in different parts of the world. He meticulously collected files and information on the numerous gay-Jewish organizations, which have formed in North America and Israel in the last two decades and enthusiastically provided his large box of files to me to aid me in my research of the subject.

Gary, a traditional Jew who was a one-time gay-Jewish commuter, proudly took on the role as one of the leaders of the organized Toronto gay Jewish community. For Gary, being involved in these activities helped to foster an integrated and secure identity:

> I was very, very involved in the gay Jewish community the first years that I was out. And that's part of my identity. I got involved in Pride Day stuff. ... I loved being in queer space and feeling safe and comfortable. I loved being the Jewish leader. I [hope] to continue to challenge Jewish institutions and families to change. I do this by being out and loud and proud as a Jew and a gay Jew as I possibly can.

Gay-Jewish integrators are made up of secular Jews as well. A secular respondent named Michael who played a leadership role in *Chutzpah* in the late 1980s had similar sentiments to share about what his involvement meant to him:

> I was coming to terms with my being both gay and Jewish at the same time. You know, you're a minority within a minority. I think that's the element of it; the need for finding people in the same category and dealing with that in a classical support-group setting was important to me. It helped me come to terms with myself, certainly. And I became certainly bolder in expressing just not the fact that I'm gay, but the fact that I'm Jewish.

4b) Challenging Jewish Theological Perspectives on Homosexuality

Another common strategy of gay-Jewish integrators is to challenge the traditional Jewish position that forbids sexual relations between those of the same gender. Eighteen respondents described this type of approach. Where one might assume it is the secular Jews who are quick to liberalize Jewish religious dictates, it is important to note that seven of the eighteen respondents in this section are traditional Jews, and of these seven, two are strictly observant. What is very interesting to observe is the distinct difference in approaches that traditional Jews take in developing more liberal perspectives in contrast to the secular Jews. As will be illustrated, traditional Jews reinterpret the Torah, while secular Jews dismiss the Torah.

Traditional Jews interviewed tend to take the position that the Torah is a divinely written document that must be respected and revered. However, the way in which the passages from the Torah that concern homosexuality (Leviticus 18:22 and 20:13) have been traditionally interpreted by religious authorities is, according to them, incorrect and not in keeping with modern times. They emphasize the dynamic and changing nature of *Halachah* (Jewish Law) and argue that more discussion and debate is necessary to develop new Jewish understandings of same-sex attraction. Danny, a religiously observant Jew, reported that he can foresee a time when Orthodox Judaism will even change its position on the issue:

> I believe that there's a way to reconcile it ... I have a gut feeling; maybe it's a hope. ... I believe that the Torah is the word of God, I'm not disputing that. I'm not saying that we sort of excise out sentences ... I think that when brilliant rabbinical minds have turned their thoughts to various issues that were important to them, they have found ways to come up with analysis that is coherent, consistent and reaches sound conclusions. And I believe that if someone were to spend his attention on the issue of prohibitions against homosexuality in the Torah and the Talmud and what it means to us today as gay people ... I think that it could be bridged.

Gary offered his opinion about the way anti-gay sentiments have been superimposed onto the Torah, while other passages in the text that support same-sex love are conveniently ignored or denied:

> Homophobia has been added into biblical stories over and over again, when in fact if you look at the biblical text, there are a lot of homo-erotic elements to it: Joseph and Potiphar story and certainly the David and Jonathan story. Those are examples of real loving relationships between men.

Despite these more liberal perspectives on Jewish Law and Jewish religious texts, traditional Jewish respondents imposed certain limits on acceptable liberal innovations. Of the seven traditional Jews who have developed more liberal interpretations of the Jewish stance on homosexuality, six of them expressed a general distaste for Reform and Reconstructionist Judaism (two denominations of Judaism in North America that are officially supportive of Jewish same-sex marriage and inclusion of open gays and lesbians in all aspects of Jewish life, including the rabbinate). Steven, a traditional respondent admitted:

> I'm still a snob about Reform. Reform [is] the movement that has opened up the most to it, obviously ... but I could just not go to a *shul* [synagogue] and see glass windows and an organ and the service in English. It makes my skin crawl. I feel like I am in church [*laughing*].

Other traditional Jews reported that they had no interest in attending religious services at *Keshet Shalom*, the (now disbanded) liberal gay Jewish congregation in Toronto, because of the non-traditional style of prayer services. Mark explained that this type of English-based service "wasn't juicy enough" and "just doesn't have any *ruach* [spirit] to me." He reported that a more traditional Hebrew service has more resonance for him. Another traditional Jew named Justin was more critical of the *Keshet Shalom* service:

> I went to that *Keshet Shalom once*, it was a novelty thing, like going to a D-list comedy theatre. It's pretty cheesy. It's just funny the way they rewrite things.

We see here a distinct style of negotiation for traditionally minded gay Jews. They tend to adopt a liberal Jewish understanding to the prohibition of homosexuality, but want to maintain their more traditional style of Jewish ritual expression that they are used to from their childhood. As Shokeid (1995:79-81) alludes to, just because one is a gay Jew does not mean one will become a Reform or Reconstructionist Jew. My data illustrate that those gay Jews who come from more traditional backgrounds have developed a more complex or nuanced method of negotiating a traditional Jewish gay identity.

Secular Jews have a different approach to gay-Jewish integration. They are not concerned with examining Jewish texts and traditions to find new approaches and interpretations that make them more supportive of homosexuality. Their sense of Jewishness contains a stronger ethnic, rather than religious, component. This means they do not need to have any investment in Jewish text or law to preserve their sense of Jewish identity. Because Jewishness can be defined as both a religious and ethnic system, it does lend itself well to this kind of strictly ethnic identification. One secular respondent named Robert, for example, explained that he feels he is "a more spiritual or cultural Jew, not a religious Jew." Zachary, elaborated well on how this type of secular Jewish identity can play out:

> At a certain point [I decided] that the Leviticus injunction against homosexuality did not mean a thing to me. I've read essays on it, but my acceptance of myself as a gay man is not going to be based on those alternative interpretations. I think some Jewish gay men need them, and so I support them. For me, they're just interesting. That means that, essentially, my view towards the Torah is that it's not the word of God. That's how I've reconciled it. No one can take your Jewish beliefs or your education or your background. It's always with you. And it's fine.

Several secular respondents offered the argument that they knew in their hearts that there is nothing wrong with them being gay. This self-realization was much more important to them than prohibitions written in a Jewish text. As such, the Torah was overruled. A secular Jew named Alan shared this opinion forcefully:

> I don't care. The Torah can say whatever it wants as far as I'm concerned. I don't follow it. I don't care about following it. Because if the Torah is going to tell me that who I am is wrong, then I'm not going to listen to it. Because I know in my heart that I'm right ... because if it wasn't right I wouldn't be feeling it.

We see here a clear example of religious individualism, Jews who feel free to appropriate only those aspects of Jewishness that they find personally meaningful.

4c) Emphasis on Linkages Between Judaism and Homosexuality

Another approach adopted by gay-Jewish integrators is the strategy of emphasizing parallels or linkages between gay and Jewish experiences. Four respondents brought forward such a perspective. Both traditional and secular respondents raised the issue of the similarities between homosexuality and Jewishness in that both involve being a member of a minority group. Kyle, a secular respondent, reported that:

> I've definitely found that the minority thing has hit. You feel like when you're growing up when your Jewish that you're a minority. And being Jewish is also like being gay in that sense. You're cognizant of being a minority, an invisible minority. I think it helps in dealing with that part of it.

This type of parallel of minority status was connected to issues of oppression and the Holocaust. A respondent named Stuart raised the issue that the Nazis in the Second World War murdered both Jews and homosexuals. Blumenfeld (1996:147-148) also raises the theme of the parallels between the dual oppressions of anti-Semitism and homophobia.

Respondents also mentioned parallels between homosexuality and Jewishness in popular culture. Harold argued that:

> There is always a link. There's the "Oy Vay" over-the-top Jewish mother and the over-the-top drag queen. Larger-than-life characters, like "the Nanny" and gay icons, Joan Rivers and people like that. There's lots of overlap. There are a disproportionate number of gay Jewish artists out there. Jews have played a huge part in defining what gay culture is. In the gay culture, there are a lot of big figures, like Tony Kushner, Harvey Fierstein; big, big names in the gay community. Lots of them are Jewish.

We also know that Jews have played a prominent leadership role in gay activism over the last several decades (Rogow 1989:79). Josh offered his insight on why this might be the case:

> Strangely enough, I think that you'll find a lot of gay Jews in the forefront of gay liberation. I think it's because Jews have already had two thousand years of being in a minority [so we] already have a lot in our genes and in our history of questioning majority establishment; we already have some mechanisms in place for questioning ultimate truth.

4d) Use of "Jewish Values" to Guide through the Gay World

Three gay-Jewish integrators who were traditional Jews discussed the idea of using their sense of "Jewish values" to help guide them in their everyday gay lives. A respondent named Charles living in one of Toronto's downtown gay neighborhoods, put it this way:

> In terms of sexual promiscuity, drug use and things that are very prevalent in the gay world, those are things where I try to draw on my Jewish values that I grew up with to help guide me and protect me. The gay world is full of temptations just like any really super-secular world is in any big city. If you live outside your ghetto, it's tough. Judaism is a little candle floating up the street, lighting my way sometimes.

When I asked Danny, an observant Jew, how he negotiates his gay and Jewish identity, he offered a similar type of perspective of the way Jewish values can inform the gay experience:

> Negotiation sounds like trade-offs from one against the other, and I don't like to think of it that way. I'd like to think that the two can in some ways enrich each other. That Jewish identity and experiences can form the way in which you should act as a gay man. Because I think that a lot of Jewish identity is centred upon a morality and a way of acting within a community, and that those are aspects of Judaism that can inform the way that you should participate in the gay community.

CONCLUSION

Due to the emphasis on "traditional" gender roles and conservative religious values, gay men and lesbians of many ethnic and religious groups feel a sense of conflict in their seemingly contradictory identities. As a result, these individuals strive to find ways to negotiate their ethno-religious and sexual identities. Recent studies on gay blacks (Icard 1985; Greene 1998), gay Latinos (Espin 1987; Garcia 1998) and gay Asians (Chan 1989) tend to simplify this question by suggesting that ethnic minority gays and lesbians *either* fully embrace their ethnic identity (by repressing their gay identity) or fully repress their ethnic identity (by embracing their gay identity). Building upon recent work on gays Christian that emphasize a larger range of strategies and the shifting nature of individually constructed identities, this study has added nuance to the discussion by highlighting the complexity and variability of this phenomenon.

As part of this larger discussion, this study has added the voices of the understudied population of gay *Jews*. Jewishness is unique among cultural identities in that it can be expressed as both an ethnicity and a religion, sometimes referred to as ethno-religious identity. This variability is shown to have important explanatory value. Borrowing from, and expanding upon, the theoretical model of Brekhus (2003), which classifies gay men into ideal types of *gay lifestylers*, *gay commuters* and *gay integrators*, I have developed new social categories to help to better understand the various negotiation strategies of gay Jewish men. In general, traditional Jews, those that tend to emphasize the religious dimension of their Jewishness, are more likely to be *Jewish lifestylers*, while secular Jews, those that tend to emphasize the ethnic dimension of Jewishness, are more likely to be *gay lifestylers*. In the case of *gay-Jewish integrators* we can find both traditional and secular Jews, but there is some variation in the methods used to achieve this integration. While both traditional and secular Jews found participation in one of the numerous Toronto gay Jewish organizations, or emphasis on the linkages between Judaism and homosexuality, to be helpful in reconciling their two identities, on the issue of challenging Jewish theological perspectives on homosexuality we see a divergence of approach. Traditional Jews tend to reinterpret Torah passages, while secular Jews dismiss what they consider outdated Torah passages. In addition, traditional Jews sometimes adapt a perspective of using what they consider the moral values of Judaism to safely guide them through their gay lives.

Beyond these general patterns is the important realization that these identities are continually negotiated and re-negotiated. Working within a theoretical framework that understands identity as a fluid journey, rather than a fixed aspect of one's life, the work has illustrated the ways that gay and Jewish identities evolve over the life course, paralleling growth and personal development. As was demonstrated, these social categories can be temporary or situational. In the case of Nathan we see a man who in his late teens and early twenties took on the extreme roles of Jewish lifestyler followed by gay lifestyler. In his thirties he moderated his approach to that of a gay-integrator. This pattern of experimentation in extremes in earlier life while seeking a more moderate strategy in later life is a common, though certainly not the only, pattern in attempting to negotiate stigmatized identities (Brekhus 2003:124-5). This leads to a discussion of age and historical cohort as important variables to consider. As opposed to respondents in their forties and fifties, respondents in their twenties or early thirties grew up in a world where being openly gay is not overly problematic. As such, these younger men saw less need for the strong identity politics of gay lifestylers or the religious insulation of Jewish lifestylers. Rather than compartmentalizing their identities (as in the case of identity commuters), or emphasizing a "master status" which defined their core self, these men had the clear expectation that multiple parts of their identities can and should be validated by society at the same time. They were thus more likely to find comfort as *gay-Jewish integrators*.

As Brekhus (2003:11 from Gamson and Moone 2004) reports in his ethnography of suburban gay men: "contrary to the public perception of a unitary, easily identifiable and coherent way to be gay (or to be any other identity), there are multiple ways to present and organize a marked identity" and "there is considerable conflict within identity categories about how to perform one's identity." This work has demonstrated that this variability of identity performance is only amplified when examining the intersections of *two* identities. Future research, which incorporates this complexity, is recommended on gay men and lesbians of other ethnic minority groups (Greeks, Italians, Chinese, etc.) as well as the significantly understudied populations of gay and lesbian Muslims, Sikhs and Hindus.

REFERENCES

Alba, R. D. 1990. *Ethnic identity: The transformation of White America.* New Haven, CT: Yale University Press.

Alpert, R. T., S. L. Elwell and S. Idelson. 2001. *Lesbian rabbis: The first generation.* New Brunswick, New Jersey: Rutgers University Press.

_____. 1997. *Like bread on the Seder plate: Jewish lesbians and the transformation of tradition.* New York: Columbia University Press.

_____. 1989. In god's image: coming to terms with Leviticus. In *Twice blessed: On being lesbian or gay and Jewish*, edited by C. Balka and A. Rose, 61-70. Boston: Beacon Press,

Balka, C. and A. Rose, eds. 1989. *Twice blessed: On being lesbian or gay and Jewish.* Boston: Beacon Press.

Barth, F., ed. 1969. *Ethnic groups and boundaries: The social organization of culture difference.* Boston: Little, Brown.

Blumenfeld. W. J. 1996. History/hysteria: Parallel representations of Jews and gays, lesbians and bisexuals. In *Queer studies: A lesbian, gay, bisexual and transgender anthology*, edited by B. Beemyn and M. Eliason, 146-162. New York: NYU Press.

Brekhus, W. 2003. *Peacocks, chameleons, centaurs: Gay suburbia and the grammar of social identity.* Chicago: University of Chicago Press.

Brodbar-Nemzer, J., S.M. Cohen, A. Reitzes, C. Shahar, and G. Tobin. 1993. An overview of the Canadian Jewish community. In *The Jews in Canada*, edited by R. J. Brym, W. Shaffir and M. Weinfeld, 39-72. Toronto: Oxford University Press.

Brown, A., ed. 2004. *Mentsch: On being Jewish and queer.* Los Angeles: Alyson Books.

Butler, J. 1990. *Gender trouble: Feminism and the subversion of identity.* New York: Routledge.

Chan, C. 1989. Issues of identity development among Asian American lesbians and gay men. *Journal of Counselling and Development* 68:16-20.

Cohen, S. M. and A.M. Eisen. 2000. *The Jew within: Self, family, and community in America*. Bloomington: Indiana University Press.

_____. 1999. Introduction. In *National variations in Jewish identity: Implications for Jewish education*, edited by S. M. Cohen and G. Horencyzk, 1-17. Albany: State University of New York Press.

_____. 1991. *Content or continuity?: Alternative bases for commitment*. New York: American Jewish Committee.

Cooper, A. 1989. No longer invisible: Gay and lesbian Jews build a movement. *Journal of Homosexuality* 18(3/4):83-94.

Davids, L. 1998. Tracking demographic assimilation. In *Jewish survival*, edited by E. Krausz and G. Tulea, 207-26. New Brunswick, N.J.: Transaction Publishers.

Dworkin, S. H. 1997. Female, lesbian and Jewish: Complex and invisible. In *Ethnic and cultural diversity among lesbians and gay men*, edited by B. Greene, 63-87. London: Sage Publications.

Eder, D., S. Staggenborg and L. Sudderth. 1995. The national women's music festival: Collective identity and diversity in a lesbian-feminist community. *Journal of Contemporary Ethnography* 23(4):485-515.

Elazar, D. J. 1999. Jewish religious, ethnic, and national identities: Convergences and Conflicts. In *National variations in Jewish identity: Implications for Jewish education*, edited by S. M. Cohen and G. Horencyzk, 35-52 Albany: State University of New York Press.

Espin, O.M. 1987. Issues of identity in the psychology of Latina lesbians. In *Lesbian psychologies: Explorations and challenge*, edited by the Boston Lesbian Psychologies Collective, 35-51. Urbana, IL: University of Illinois Press.

Fink, A. S. and J. Press. 1999. *Independence park: The lives of gay men in Israel*. Stanford, California: Stanford University Press.

Fishman, S.B. 2000. *Jewish life and American culture*. Albany, New York: State University of New York Press.

Foucault, M. 1980. *The history of sexuality, volume 1*. Translated by Robert Hurley. New York: Vintage.

Gamson, J. and D. Moon. 2004. The sociology of sexualities: Queer and beyond. *Annual Review of Sociology* 30:47-64.

Gans, H. J. 1979. Symbolic ethnicity: The future of ethnic groups and cultures in America. *Ethnic and Racial Studies* 2(1):1-20.

Garcia, B. 1998. *The development of Latino gay identity*. New York: Garland Publishing, Inc.

Goffman, E. 1963. *Stigma: Notes on the management of spoiled identity*. Englewood Cliffs, N.J: Prentice-Hall, Inc.

Greenberg, S. 2004. *Wrestling with god and men: Homosexuality in the Jewish tradition*. Madison, Wisconsin: University of Wisconsin Press.

Greene, B. 1998. Family, ethnic identity and sexual orientation: African American lesbians and gay men. In *Lesbian, gay and bisexual identities in families: Psychological perspectives*, edited by C. J. Patterson and A. D'Augelli, 40-52. London: Oxford University Press.

Horowitz, B. 2002. Reframing the study of contemporary Jewish identity. *Contemporary Jewry* 23:14-34.

_____. 1998. Connections and journeys: Studying identities among American Jews. *Contemporary Jewry* 19:63-94.

Icard, L. 1985. Black gay men and conflicting social identities: Sexual orientation versus racial identity. *Journal of Social Work and Human Sexuality* 4(1/2):83-93.

Levado, Y. 1993. Gayness and god: Wrestlings of an orthodox rabbi. *Tikkun* 8(5):54-60.

Mahaffy, K. A. 1996. Cognitive dissonance and its resolution: A study of lesbian Christians. *Journal for the Scientific Study of Religion* 35(4):392-402.

Moore, T., ed. 1995. *Lesbiot: Israeli lesbians talk about sexuality, feminism, Judaism and their lives*. New York: Cassell.

Morales, E. 1990. Ethnic minority families and minority gays and lesbians. *Marriage and Family Review* 14:217-239.

Mushkat, D. 1999. 'Alienated Jews': What about outreach to Jewish lesbians? *Journal of Jewish Communal Service* 75(4):239-247.

Nagel, J. 2000. Ethnicity and sexuality. *Annual Review of Sociology* 26:107-33.

_____. 1994. Constructing ethnicity: Creating and recreating ethnic identity and culture. *Social Problems* 41(1):152-76.

Raphael, L. 1996. *Journeys and arrivals: On being gay and Jewish*. Boston: Faber and Faber.

Rapoport, C. 2004. *Judaism and homosexuality: An authentic Orthodox view*. London: Vallentine Mitchell.

Rodriguez, E. M. and S.C. Ouellette. 2000. Gay and lesbian Christians: Homosexual and religious identity integration in the members and participants of a gay-positive church. *Journal for the Scientific Study of Religion* 39(3):333-47.

Rogow, F. 1989. Speaking the unspeakable: Gays, Jews and historical inquiry. In *Twice blessed: On being lesbian or gay and Jewish*, edited by C. Balka and A. Rose, 72-84. Boston: Beacon Press.

Roof, W. C. 1999. *Spiritual marketplace: Baby boomers and the remaking of American religion*. Princeton, N.J.: Princeton University Press.

Rose, A. 1999. Deconstructing Leviticus: Finding a queer spiritual path back from the exile of a sacred text. In *ReCreations: Religion and spirituality in the lives of queer people*, edited by C. Lake and M. Drache, 117-124. Toronto: Queer Press.

Schimel, L. 2000. *Kosher meat*. Santa Fe, New Mexico: Sherman Asher Publishing.

Schnoor, R. F. 2003. *Finding one's place: Ethnic identity construction among gay Jewish men*. Doctoral Dissertation. Montreal: McGill University, Department of Sociology.

Sedgwick, E. K. 1990. *Epistemology of the closet*. Berkeley: University of California Press.

Shaffir, W. and M. Weinfeld. 1981. Canada and the Jews: An introduction. In *The Canadian Jewish mosaic*, edited by M. Weinfeld, W. Shaffir and I. Cotler, 7-20. Toronto: John Wiley and Sons.

Shneer, D. and C. Aviv., eds. 2002. *Queer Jews*. New York: Routledge.

Shokeid, M. 1995. *A gay synagogue in New York*. New York: Columbia University Press.

Spector, M. and J. I. Kitsuse. 1987. *Constructing social problems*. New York: Aldin de Gruyter.

Thumma, S. 1991. Negotiating a religious identity: The case of the gay evangelical. *Sociological Analysis* 52(2):333-47.

Torton Beck, E., ed. 1989. *Nice Jewish girls: A lesbian anthology*. Boston: Beacon Press.

Tulchinsky, K. X., ed. 2000. *Friday the rabbi wore lace*. San Francisco, CA.: Cleis Press.

Walzer, L. 2000. *Between Sodom and Eden: A gay journey through today's changing Israel*. New York: Columbia University Press.

Waters, M. C. 1990. *Ethnic options: Choosing identities in America*. Berkeley and Los Angeles, CA: University of California Press.

Wilcox, M. M. 2003. *Coming out in Christianity: Religion, identity and community*. Bloomington: Indiana University Press.

_____. 2002. When Sheila's a lesbian: Religious individualism among lesbian, gay, bisexual and transgender Christians. *Sociology of Religion* 63(4):497-513.

wolfman, o. 2002. Remaking family: Canadian Jews, sexuality and relationships. In *Queer Jews*, edited by D. Shneer and C. Aviv, 156-171, New York: Routledge.

Wuthnow, R. 1998. *After heaven: Spirituality in America since the 1950s*. Berkeley: University of California Press.

Gay and Lesbian Christians: Homosexual and Religious Identity Integration in the Members and Participants of a Gay-Positive Church

ERIC M. RODRIGUEZ, M.A.
SUZANNE C. OUELLETTE, PH.D.

In this study we explore individual's experiences of identity integration between their sexual orientation and religious beliefs. Using both qualitative and quantitative research methods, we examined identity integration in forty members and participants of the Metropolitan Community Church of New York (MCC/NY), a gay-positive church located in Mid-town Manhattan. The survey and interview data collected showed that: (1) a majority of the research participants reported that they had successfully integrated their homosexual and religious identities, (2) being integrated was related to higher role involvement at MCC/NY, being a member of the church, attending more MCC/NY worship services and activities/ministries, and attending MCC/NY for more years, (3) lesbians were less likely than gay men to report past conflict between their identities, and more likely to report being fully integrated, and (4) MCC/NY played an important role in helping these participants achieve integration between their homosexual and religious identities.

"Sure, I mean [being gay and Christian] is the big thing that religious gay people grapple with isn't it? There's homophobia in there, and there's fear of divine retribution, there's all of those things. What if I'm wrong? What if there's a Hell and I'm going there because I'm a faggot, and I have sex with men?"

"Being lesbian and Christian is accepting the unique creation that God created me to be. Love is beautiful no matter how it's expressed. Difference doesn't have to mean sin. The Bible has been used against us and it really isn't all about what's been presented to us. The Bible is not the book that hits us in the head, it's the book that leads us into light."

These reflections from two Christian church participants reveal the complexity and diversity in the experiences of people who define themselves as both gay and religious. Much of the research on gay and lesbian people of faith assumes that any reconciliation of religious and homosexual identities is preceded by a history of conflict (Thumma, 1991). Investigators link this conflict with the fact that although a few Christian denominations and denominational religious groups, such as the United Church of Christ, Integrity in the Episcopal Church, Dignity in the Roman Catholic Church, and Lutherans Concerned, view homosexuality in a more tolerant light, most mainstream Christian denominations do not (Ellison, 1993; Mahaffy, 1996). In one study, 72% of surveyed churches and organizations condemned homosexuals and homosexuality as being an abomination in the eyes of God (Melton, 1991). Members and participants in many of the more conservative Christian denominations refer to gays and lesbians as "unnatural", "evil", "sinners", and "perverts" (Clark, Brown, and Hochstein, 1990; Greenberg and Bystryn, 1982; Keysor, 1979; Scanzoni and Mollenkott, 1978). When such anti-gay language and sentiment is encountered by gays and lesbians with a strong religious faith and a strong positive feeling towards their sexual

---

Eric M. Rodriguez is a student in the Doctoral Program in Psychology, Subprogram in Social/Personality Psychology, at The Graduate School and University Center of The City University of New York, 365 5th Avenue, New York, NY, 10016-4309. E-mail: eric_m_rodriguez@hotmail.com.
Suzanne C. Ouellette is a professor in the Doctoral Program in Psychology, Subprogram in Social/Personality Psychology, at The Graduate School and University Center of The City University of New York.

orientation, they enter a situation where identity conflict can occur. Indeed, researchers have applied the label of cognitive dissonance to the struggle they have documented between homosexuality and religious beliefs among lesbians (Mahaffy, 1996) and gay men (Thumma, 1991).

As the opening quotations reveal, there are strikingly different ways in which individuals cope with situations that present them with interpersonal, social, and ideological forces so much at odds with their own experience of their identities. Research of both gay and lesbian and general identity processes reveals four general categories of identity work in which gays and lesbians might engage to alleviate conflict between their homosexual and religious identities.

STRATEGIES FOR DEALING WITH BEING BOTH HOMOSEXUAL AND RELIGIOUS

Rejecting the religious identity. The first strategy is a matter of divorcing oneself from the Christian religion. Research has shown that as many as 62% of gays and lesbians feel that religion is not an important aspect of their lives (Singer and Deschamps, 1994). Rejecting Christian beliefs is accomplished by becoming an atheist, or by becoming involved in a non-Christian religion that does not hold negative views towards homosexuals and homosexuality (Ellison, 1993). While formal steps such as these appear to be fairly common in the gay and lesbian community, rejection of one's Christian beliefs can also be a very subtle experience: people no longer attend any type of worship service, they no longer pray or make any references to the presence of God and/or Christ. They simply allow their religion to slip quietly out of their lives.

Rejecting the homosexual identity. In the second strategy, the person with the homosexual identity seeks to become a heterosexual, most commonly through reparative or Christian "conversion" therapy (Piazza, 1994). Whether or not such therapy actually works has been the subject of much debate (see Keysor, 1979; Scanzoni and Mollenkott, 1978; Throckmorton, 1998). However, proponents of Christian conversion therapy maintain that it is a valid way of ridding oneself of homosexual longings and reaffirming one's love for God, Christ, and Church (Keysor, 1979). Rejecting the homosexual identity can also be accomplished by sexual abstinence on the part of the gay or lesbian person. In some Christian denominations, as long as the person is not engaging in homosexual behavior then they are not considered to be a homosexual (Keysor, 1979; Thumma, 1991).

Compartmentalization. Psychological investigators of general identity processes have identified yet a third strategy for eliminating conflict. According to Baumeister, Shapiro, and Tice (1985), compartmentalization is a compromise between conflicting identities. By keeping two conflicting identities separate, conflict resolution, or identity consonance, is achieved. Baumeister et al. state: "[Compartmentalization] requires that the two spheres be kept rigidly separate in order to avoid conflicting prescriptions for behavior" (Baumeister et al., 1985). Gays and lesbians can use this strategy by keeping their religion out of the homosexual parts of their lives, and keeping their homosexuality out of their religious lives. However, only by completely isolating one's homosexual identity from one's religious identity can this strategy for reducing identity conflict be successful. If the barriers between the two identities are breached, a slide back to identity dissonance could result.

Identity integration. The focus of the current study is on this fourth strategy. Deaux (1991) observes that identities can change over time and persons can respond to identity threats through the addition of a new identity. Accordingly, we propose that identity conflict can be alleviated when gay men and lesbians integrate their religious beliefs and their homosexuality into a single, new, workable understanding of the self. Such individuals hold a positive gay identity, a positive religious identity, and do not feel conflict between the two. To differentiate integration from compartmentalization, we assume that gays and lesbians who experience integration combine their two identities rather than keeping them in separate spheres of their life. Such people have no self-imposed walls between their homosexuality and their religious beliefs, and perceive societally-imposed barriers as surmountable. This creates a new, complex and yet coherent identity: Gay or lesbian Christian.

SITES OF HOMOSEXUAL AND CHRISTIAN IDENTITY INTEGRATION

Key forums for homosexual and Christian identity integration are groups that promote both homosexuality and religion. Thumma (1991) depicts such an assembly in his research on gay men attending the Evangelical group Good News; and Wagner, Serafini, Rabkin, Remien, and Williams (1994) in a study of gay men attending the gay Catholic organization Dignity. The goal of these religious support groups to deliver both gay-positive and Christian-positive messages has been made easier by the recent emergence of a "gay theology" that specifically values gay men and lesbians of the Christian faith, and recognizes their spiritual needs. These groups have re-interpreted the Bible in such a way that homosexuality is viewed in a positive, rather than negative, religious light (Englund, 1991; Thumma, 1991). As a result, many of the gay and lesbian people of the Christian faith involved in such organizations have begun to hear the message that God loves all people, including homosexuals. Gay and religious-positive support groups such as Good News and Dignity are, however, typically affiliated with Christian churches that view homosexual behavior as a sin.

The present research builds on examinations of gay and lesbian Christian support groups by studying a gay-positive church - a formal Christian religious institution that preaches a positive message about homosexuality and ministers specifically to the gay and lesbian community. It is important to note here that a *gay-positive* church should not be confused with a *gay-friendly* church. A gay-friendly church may welcome (either implicitly or explicitly) the participation of gays and lesbians, but such churches do not typically address the religious and spiritual needs inherent in the gay and lesbian community. Becoming increasingly involved in a gay-positive church, according to leaders of such churches, enables gays and lesbians to alleviate the conflict between their religious beliefs and their homosexuality while increasingly enjoying identity integration (Lukenbill, 1998; Perry, 1990; Piazza, 1994; White, 1994).

In previous research (Rodriguez and Ouellette, 1999), we conducted a participant-observation study of The Metropolitan Community Church of New York (MCC/NY), a gay-positive Christian church that ministers primarily to the gay, lesbian, bisexual, and transgendered community in New York City. MCC/NY's new member brochure states that its members are "committed to exploring and deepening their spirituality, and building a loving, supportive, and diverse community" (MCC/NY, 1994). This particular gay-positive church is one of over 300 churches in the Universal Fellowship of Metropolitan Community Churches (UFMCC). UFMCC claims to be the world's largest and oldest gay, lesbian, and bisexual organization, and also claims to be the fastest growing religious organization in the world (MCC/NY, 1994). Founded by the Reverend Doctor Troy Perry in 1968 in Los Angeles, California, UFMCC has since extended across the United States and to 16 other countries (Lukenbill, 1998; Singer and Deschamps, 1994).

At the time of the participant-observation study, MCC/NY had a mailing list of 1500 people, with approximately 170 members and an average weekly church attendance of nearly 200 people (MCC/NY, 1994). The Pastor of the church is a self-identified lesbian who has been Pastor at MCC/NY for over 10 years. While the majority of MCC/NY's church attendees are gay or lesbian, some heterosexual, bisexual, and transgendered individuals also attend. The church is ethnically diverse with approximately 60% White, 15% Black, 15% Hispanic and 10% Asian, while the gender makeup of the church is equally divided between males and females (Rodriguez and Ouellette, 1999). In addition to two Sunday church services and a weekly Bible study class, MCC/NY has a number of other ministries as well (e.g., Young Adult Ministry, HIV Support Group, Hispanic Outreach, Sunday School, etc.). While MCC/NY's ministries are primarily for the benefit of gays and lesbians, several of them (e.g. a Food Pantry and a Homeless Shelter) extend beyond the gay and lesbian community.

In our participant-observation study of MCC/NY worship services, Bible study classes, and MCC/NY literature, we uncovered several specific strategies used by the church to facilitate the creation of a combined gay or lesbian Christian identity. In addition to documenting MCC/NY's use of inclusive and gender neutral language, as well as the use of the historical-critical method of interpreting the Bible, we: (1) described how the structure of the liturgy at MCC/NY enabled its members and participants to recognize the church as a legitimate religious institution that is simultaneously gay/lesbian and Christian, (2) characterized how it was that MCC/NY, through the preaching of its lesbian Pastor, provided gays and lesbians with a positive way of thinking about themselves as gay and lesbian Christians, and (3) documented a struggle that this church shares with every other moral community; namely the challenge to provide one group of people with a strong sense of who they are as a valued group without creating boundaries that exclude others (Rodriguez and Ouellette, 1999).

THE CURRENT STUDY

The study we present here sought to examine the link between extent of involvement in MCC/NY and identity integration, and to better understand homosexual and religious identity integration as perceived by the members and participants attending MCC/NY. While there are questions about homosexual and religious identity in a variety of Christian denominations, as well as other religious faiths (i.e. Islam, Judaism, and non-monotheistic religions such as Buddhism and Hinduism), for this work we decided to concentrate on one particular gay-positive Christian church, and on the gay and lesbian members and participants of that church.

We report our results in the order of the increasing complexity of our research questions and hypotheses. We begin with our test of the hypothesis that the more participants are involved with MCC/NY, with regard to role involvement, worship service attendance, church activity/ministry attendance, formal membership, and number of years at MCC/NY, the more likely they are to report integration between their religious and homosexual identities. We then move to ask how integration is related to other characteristics of MCC/NY participants, namely openness about their homosexuality, personal religious history, age, and gender. Finally, we look more closely at integration itself as we ask under what circumstances research participants were aware of themselves as integrated, and what they felt caused their homosexual and Christian identities to integrate. Here we test the hypothesis that participants would report MCC/NY as playing a major role in their identity integration.

METHOD

Procedure

The first author went to both morning and evening services at MCC/NY on six consecutive Sundays, and gave a brief statement about the project during the announcements portion of the worship service to recruit participants. The congregation was informed of the first author's involvement as a member of MCC/NY for over a year prior to the start of the research. We relied upon the first author's "insider status" (Jorgenson, 1989; Marshall and Rossman, 1989) as a study strength that encouraged individuals with differing levels of experience at MCC/NY to participate in the study. The first author met with potential participants in the church social hall immediately following worship services. During that meeting the first author informed each volunteer that he or she would be asked, at a time and place of their choice, to sign a consent form, fill out a brief survey, and participate in a tape-recorded, semi-structured interview. Only 6 of the 40 interviews (15%) took place at the church itself. The rest took place in a non-church related location (i.e. the participant's home, the interviewer's home, a restaurant, etc.). Analyses with location as the grouping variable showed that the setting of the interview did not impact any of the results. The average taped interview was 78 minutes, ranging from 30 minutes to 3 hours.

Table 1
Self-Reported General and Religious Demographics

VARIABLES	N	%	VARIABLES	N	%
Sexual Orientation			Integration		
-Gay Male	22	55%	-Not Integrated	10	25%
-Lesbian	18	45%	-Integrated	29	73%
Race/Ethnicity			-Missing	1	2%
-White	26	65%			
-Hispanic	5	12.5%	Religious History		
-Black	4	10%	-Catholic	13	33%
-Asian	3	7.5%	*Protestant	27	67%
-Am. Indian	2	5%	-Evangelical	12	30%
Member of MCC/NY			-Nonevangelical	28	70%
-Yes	22	55%	-No Past Conflict	12	30%
-No	18	45%	-Past Conflict	28	70%

VARIABLES	MEAN	STANDARD DEVIATION	RANGE	POTENTIAL RANGE
Age	37.35	9.76	21 – 64	NA
Openness	7.73	1.47	4 – 9	0 – 9
-Extended Family	5.78	3.32	0 – 9	0 – 9
-Immediate Family	7.78	2.32	0 – 9	0 – 9
-Friends	8.83	.50	7 – 9	0 – 9
-School	8.30	1.64	4 – 9	0 – 9
-Work	7.49	2.27	0 – 9	0 – 9
Years at MCC/NY	4.60	4.88	0 – 17	NA
Role Involvement	8.38	6.46	0 – 21	0 – 25
MCC/NY Attendance				
-Worship Services	4.10	1.58	1 – 6	1 – 6
-Ministries/Activities	3.08	1.58	1 – 6	1 – 6

*Protestant Denominations included Anglican (N=2), Baptist (6), Christian Scientist (1), Congregational (2), Dutch Reformed (1), Episcopalian (1), Evangelical (1), Fundamentalist (1), General Protestant (4), Lutheran (2), Methodist (4), Pentecostal (1), and Presbyterian (1).

Participants

Because our aim was to conduct in-depth, semi-structured interviews along with collecting survey data, we chose to suspend participant recruitment after acquiring 41 volunteers. We felt that given the exploratory nature of the research, 41 participants was a manageable number to use for the qualitative analyses and for running simple statistical tests. Of the 41 volunteers only one person dropped out, leaving a final research sample of 40 individuals (22 males and 18 females). The participants' self-reported demographic and religious information is presented in Table 1.

While random sampling was never a viable option for this study because of the inherent difficulties in conducting research with gays and lesbians (see Gonsiorek, 1991), and the level of personal interaction required by the qualitative aspects of our study protocol, we achieved a sample that accurately represented the population of MCC/NY. The gender and racial breakdowns of the research sample corresponded closely with the actual percentages found in the participant-observation study conducted previously (Rodriguez and Ouellette, 1999). Additionally, the variance found in membership, length of time at MCC/NY, and worship service and activity/ministry attendance matched that of the entire congregation of MCC/NY. Due to our use of non-probability sampling, care has been taken to insure that any generalizations made regarding this research apply only to the members and participants of MCC/NY.

Materials

Both the survey and the interview protocol were designed by the authors. The first included questions about participants' history at MCC/NY and general and religious demographic information. The in-depth, semi-structured interviews were guided by 22 questions, crafted through the help of our conceptual framework and 5 pilot interviews with MCC/NY participants. In the interview, participants spoke of their personal experiences at MCC/NY, how they defined being gay verses being religious and/or spiritual, their religious life histories, and the interaction of their homosexual and religious identities (Rodriguez, 1997).

Reported in this article is the analysis of transcribed responses to questions from the final section of the interview that dealt specifically with the issue of identity integration. This is where we heard participants summing up their self-understanding of, and where we found the richest narratives on, their current views of their homosexual and Christian identities (see Rodriguez and Ouellette, 2000, for work we have conducted using the interviews in their entirety). The key thematic questions that shaped this section of the interview were: (1) Right now as you are sitting there, can you describe for me what it's like being both gay or lesbian and Christian? (2) Have there ever been times when you've been both gay or lesbian and Christian at the same time? Tell me about those times. (3) Today, do you feel that you have combined together your sexual orientation and your religious beliefs? What caused them to combine or not?

In the spirit of interpretive social science (Rosenwald and Ochberg, 1992), we understood what the research participants said in the interviews not to be an objective revelation of "what really happened," but an example of their continuing narrative efforts to make sense of who they are as gay and lesbian Christians. The interviewees, who were informed of the first author's involvement at MCCNY and of the goals of the research, understood the "shared task and purpose" of the interviews to be a representation of how they struggled with being gay or lesbian and religious (Mishler, 1990). We viewed the interviews as texts that enabled dialogue to take place both between ourselves as researchers, and between ourselves and the participants (Smith, 1993). Thus the interviews were opportunities for us, as well as for each of the gay and lesbian research participants, to delve more deeply into their gay and religious lives.

Quantitative Analyses

Frequencies, descriptives, cross-tabs, and one-way ANOVAs were conducted using the following variables. We present descriptive information for these variables in Table 1.

Integration. Although the interviews revealed integration to be a long-term, ongoing, and complex process, coders were able to reliably assess the extent to which participants currently view their homosexual and religious identities as integrated. Using a 3-point Likert scale, they assigned a single score of 0 to indicate not at all integrated, 1 for somewhat integrated/working on it, and 2 for fully integrated to each transcript that they read. The inter-rater reliability between the first author and an outside coder unaware of the study hypotheses was .87. To obtain adequate sample sizes for each category, integration was subsequently recoded into a dichotomous "fully integrated" (1) verses "not fully integrated" (0) variable.

Involvement. A measure of *role involvement* was taken directly from close-ended items in the survey. Participants were read a series of roles that occur at MCC/NY (i.e., usher, lay reader, choir member, etc.), and asked if they had ever filled one of these roles. Participants were also read a list of activities and ministries that take place at MCC/NY (i.e. Bible study, HIV/AIDS support group, Hispanic Outreach, etc.), and asked if they have ever been involved in any of these. Participants received one point for each role and activity/ministry that they had ever been involved in, with duplicate items (i.e. choir member as both a role and an activity) counted only once. Each participant's responses were summed, leading to an involvement score with a range of 0 - 25 (0 = no involvement). Using Cronbach's Alpha, reliability for this scale was estimated at .89. Other involvement measures taken directly from the survey included *membership* (0 = not a member, and 1 = an official member of MCC/NY), *number of years attended MCC/NY, worship service attendance* (six-point Likert scale with 1 = attended worship services less than five times per year, and 6 = attended worship services twice a week), and *attending activities/ministries* (six-point Likert scale with 1 = did not attend church activities/ministries, and 6 = attended activities/ministries 2 or more times per week).

Religious History. Participants reported in the survey what religion they considered themselves when growing up (see Table 1). For analysis purposes the resulting responses were then coded into two separate dichotomous variables: Protestant (1) verses Catholic (2), and evangelical (1) verses nonevangelical (2). The evangelical/ nonevangelical dichotomy was based on a categorization developed by Jelen and Wilcox (1991), and utilized by Mahaffy (1996) to assess religious history in her study of lesbian Christians. We used *past conflict* as an additional measure of perceived religious history. Participants were asked during the interview whether or not they had ever experienced conflict between their sexual orientation and their religious beliefs. We developed a dichotomous coding scheme for this construct, with 0 = never experienced conflict, and 1 = have experienced conflict. The inter-rater reliability between the first author and an outside coder unaware of the study hypotheses was .98.

Other Variables. Some of the other variables of interest included participant's *age*, and *openness about their sexual orientation*. We measured openness as a series of six ten-point (0-9) Likert scale items: general openness, openness with immediate and extended family, friends, at school, and at work.

Qualitative Analyses

Participants' interview transcripts were reviewed with structured coding procedures (as described above) and through qualitative thematic and content analyses. Following several readings of the narratives, thematic analysis (Smith, 1993) enabled us to select specific quotations that most strongly reflected critical identity themes, and represented the racial, ethnic, and gender diversity of participants. Content analysis (Marshal and Rossman, 1989) was conducted on the narratives from participants who were assigned a "fully integrated code," with the aim of establishing a list of reasons offered for integration. We used the resulting frequency distribution to test the hypothesis that participants would report that MCC/NY played a major role in their identity integration. An additional thematic analysis was conducted to select the quotations that best illustrated the range of respondent views on the specific role MCC/NY played in helping individuals integrate their homosexual and Christian religious identities. Please note that pseudonyms were assigned to each participant, and any potentially identifying information was either edited or deleted from the transcripts to protect the confidentiality of the research participants.

Results

Integration and Involvement with MCC/NY

Twenty-five percent (N = 10) of the forty research participants reported not being fully integrated, while 72.5% (N = 29) reported being fully integrated (one person was coded as missing - see Table 1); indicating a considerably integrated research sample. Miriam, a 29 year old White lesbian who was a member of MCC/NY and had attended for less than 3 years, provided an example of a participant who reported that she had integrated her homosexual and Christian religious identities. She stated:

> *I can say with conviction and with real faith that I know I am a child of God. I am loved by God, being gay is the way I was created. I feel strongly that if I fall in love with somebody and were to have a holy union that it would be blessed by God. My sexuality is very much part of my religion.*

Isaac, a 23 year old Black gay man who was not a member of MCC/NY and had been attending for 2.5 years, provided an example of a participant who was coded as not being fully integrated. Isaac stated:

> *It's like I believe in God and everything but I still want to do the stuff that I'm doing now. So it's not combining. My parents put it to me this way – 'If you want to be gay, be gay. If you're going to be Christian, you can't do gay sexual acts and be a Christian at the same time, it's not right. You can't combine both of them at the same time.'*

While the term "integration" was not used by the interviewer during the course of the interview, about 10% (N = 3) of the integrated research participants used the word "integrated" on their own when describing how they have combined their sexual orientation and their religious beliefs. Additional synonymous terms that were utilized by both the integrated and non-integrated research participants included: entangled, melded, reconciled, combined, congruence, fusion, together, and wedded. Thus the concept of integration as a representation of a distinct identity that combines together one's sexual orientation and religious beliefs was quite salient in many of the participant's minds.

Cross-tab analyses on the dichotomous integration variable (fully integrated verses not fully integrated) revealed that members were significantly more likely to report being fully integrated than non-members (86.4% verses 58.8%; $c^2(1) = 3.82$, $p < .05$). One-way ANOVAs were conducted using fully integrated as the grouping variable on the dependent variables of involvement, attending church ministries/activities, attending worship services, and number of years at MCC/NY. Being fully integrated was significantly related to a higher role involvement ($F(1,37) = 5.45$, $p < .03$), higher worship service attendance ($F(1,37) = 12.98$, $p < .001$), and a higher ministry/activity attendance ($F(1,37) = 5.66$, $p < .02$). Integration was also significantly related to the number of years a gay or lesbian participant attended MCC/NY, although this finding was only at the trend level of significance ($F(1,37) = 2.91$, $p < .09$). These findings support our hypothesis that the more individuals are involved with MCC/NY, the more likely they are to report integration between their religious and homosexual identities.

Integration and Openness, Age, Religious History, and Gender

Participants who reported that they had fully integrated their sexual orientation and their religious beliefs were found to be significantly more open about their sexual orientation in general ($F(1,37) = 4.12$, $p < .05$), and were also more open about their sexual orientation at work ($F(1,36) = 5.80$, $p < .02$). Participants who claimed full integration were also found to be older ($F(1,37) = 3.34$, $p < .08$), although this finding was only at the trend level of significance. With regard to religious history, there was a notable lack of significant results. No relationship was found between integration and whether one was raised Protestant or Catholic, or in an evangelical rather than nonevangelical setting. Also, no association was observed between being fully integrated and perceived experiences of past conflict.

With regard to gender, a cross-tab analysis indicated a trend level result where lesbians were more likely than gay men to report being fully integrated (88.2% verses 63.6%; $c^2(1) = 3.04$, $p <$

.08). As we analyzed interview responses for quotations that would best represent the integration themes, we found the majority of such quotations from women, while the men provided essentially all of the telling lack of integration statements. Note the differences between the following two quotations. Ruth, a 47 year old Black lesbian who was a member of the church and had been attending for just under 6 years, stated:

Lesbian and Christian, it's like breathing, they work together, there's nothing separate about it. I love my spouse, but if I'm walking down the street and I see somebody who needs my help, if I can possibly help them I will help them. That's being Christian. It doesn't take away from the fact that I'm also a lesbian.

Jacob, on the other hand, a 40 year old White gay man who was a member of the church and had been attending for more than 12 years stated:

It isn't just complete. This is not over, this is a struggle I still have to fight. And to keep those two things where they belong. In other words, God's view is that 'yes they are the same thing,' my view, 'yeah they are the same thing but I don't always believe it.'

Struck by these gender differences, we ran a set of post-hoc exploratory analyses to determine if there were other differences between the gay and lesbian participants. While one-way ANOVAs did not reveal a gender difference in regards to role involvement, membership, and number of years at the church, the women attended significantly more MCC/NY church activities/ministries ($F(1,38) = 4.09$, $p < .05$) and also had a tendency to attend more worship services than the gay men ($F(1,38) = 3.35$, $p < .08$). Also, in a cross-tab analysis, we found that gay men were significantly more likely than lesbians to report having experienced conflict between their homosexuality and religious beliefs (90.9% verses 44.4%; $c^2(1) = 10.18$, $p < .001$). All but two of the gay men told stories of conflict while less than half of the lesbians did.

Situated Awareness of Both Identities

Regardless of gender, both the integrated gay men and the integrated lesbians discussed situations where they were aware of being both homosexual and Christian at the same time. In response to the question "Have there ever been times when you've been both gay/lesbian and Christian at the same time?", many of the participants reported that they were "always" aware of being both throughout their entire lives, regardless of the situation. Miriam, the 29 year old White lesbian we met earlier, illustrated how she had "always" been aware of both her lesbian and Christian identities:

I've never stopped holding onto my belief in God. There have been times I've been angry with God. There are times when I've felt a little cut off, but I've been gay from the time I rocked into this world, and I've been a child of God from the time I rocked into this world. I've always believed in God.

Unlike Miriam, several participants mentioned that such a sense of awareness had only come about recently in their lives; within the past year or so for such individuals. Timothy, a 33 year old White gay male who was a member of MCC/NY and had been attending for four years, said:

It's a real recent thing, the past couple of years. The religious right forced me to come up with a way to justify my life [as a gay Christian] in a real way, not a cosmetic way but in a way that's unified, that both go together and equal the whole that is me. As I said it's recent, and it was a struggle, but I've had to do it for my own self-esteem.

For Timothy, as well as for other participants, MCC/NY was mentioned as the place where they felt most aware of being both gay or lesbian and Christian. In addition to MCC/NY, however, several participants also mentioned other specific situations where their awareness of both identities was most salient. We use the following three quotes to illustrate this point. Sarah, a 43 year old White, lesbian, member of MCC/NY who had been attending for over four years, stated:

> *I suppose I'm most aware of both at [MCC/NY], especially when I'm listening to the Pastor. Also when I'm thinking about issues related to the oppression of lesbians. I guess when I'm attacked by other Christians, I'm aware of 'you're attacking me and I'm one of you, I'm Christian'. It's in those kinds of political contexts that I'm aware of both.*

Luke, a 26 year old White gay man who was a member of the church and had been attending for over a year, echoed Sarah's sentiments about MCC/NY, but also admitted that he had never thought about this issue before. Luke stated:

> *I've never thought of that, other than inside MCC/NY, which is probably the only place you're ever thinking about both at the same time. I'd also say [marching with MCC/NY] at the [gay] pride parade last Sunday. [At the parade] I felt a very overwhelming presence of being both and being very fully both, and that's probably the first time I felt that rush.*

For Sarah the "other situations" included any time when she was being confronted by individuals or institutions representing the anti-gay, conservative Christian view point. For Luke the "other situation" included marching with MCC/NY in the New York City gay pride parade. Joseph, a 27 year old White gay man who was a member of the church and had been attending for three years, also mentioned marching in the gay pride parade as a situation where both his gay and Christian identities were salient. He stated:

> *I mean, marching in the gay pride parade, just walking down 5th avenue, ministering to other gay men, and showing up at [MCC/NY] and giving communion in drag. I mean, how gay can you get. It's to celebrate with who I am with all I am in front of God. It's been wonderful.*

While participants described other situations where they were aware of being both gay and Christian (political contexts, gay pride parades; one person even mentioned attending another church with their parents), a majority mentioned their involvement with the gay and Christian-positive MCC/NY as the situation where both identities were most salient. This occurred both within the confines of the church itself, as well as outside the church as MCC/NY took its message to the streets of New York.

The Impact of MCC/NY

Those participants who reported that they had fully integrated their homosexuality and their Christian religious beliefs were asked what caused these two identities to combine. Table 2 illustrates the variety of reasons participants cited as causing their identity integration. While issues of God and spirituality did play an important role for several individuals, most of the fully integrated participants used more concrete strategies such as becoming involved in MCC/NY, reading relevant literature, and simply talking with others about their concerns. For many of the gay and lesbian participants in this study, simply having a strong Christian faith was not always enough - an additional, more tangible, step was often necessary for identity integration to occur.

Of all the reasons cited, involvement in MCC/NY was far and away the most frequently described. All but five of those participants who reported identity integration claimed that MCC/NY helped them to realize and maintain a sense of themselves as both gay/lesbian and Christian. Samuel, a 37 year old Hispanic gay man who was a member of MCC/NY and had been attending for 8 years, puts this well for the group:

> *MCC[NY] has been a major participant, or player, in this process. MCC[NY] has provided me with the tools to enhance not only my spiritual life but also my life as a homosexual person. And [MCC/NY] has helped me to develop ways to integrate them in a healthy and...intelligent way.*

Two other participants speak of the special sort of collective or community that MCC/NY creates, one that heals and that is also different from their previous experiences. Mary, a 35 year

Table 2
Self-Reported Causes of Identity Integration (N=29)

REASONS STATED	N	%*
MCC/NY	24	83%
Knowledge, Reading, Education	8	31%
Accepting Self, Sense of Completeness	8	31%
Spiritual Reasons, The Work of God	7	24%
Talking to Others	5	17%
Maturity	2	7%
HIV/AIDS	1	4%
The Religious Right	1	4%
Other/Unknown	1	4%

*Column does not add up to 100% because participants reported more than one reason for integration.

old Asian-American lesbian who was a member of MCC/NY and had been attending the church for just over a year, said:

> [MCC/NY] made it okay for me to be lesbian and Christian. You know, the outlet is there. And I think just dealing with other people of faith that have had the same struggle, you know, knowing that it's okay to go through all of those things.

And for Thomas, a 38 year old White gay man who was a member of the church and had been attending for 3.5 years:

> There's not much out there that is life affirming, and embraces sexuality at the same time. It is so liberating and refreshing to be with men and not know what they look like naked. The main reason we get together is because of God and our beliefs, not because we're pretty, or cute, or big and muscular, or whatever.

In participants' narratives, MCC/NY emerges as a very special, indeed sacred, space for identity work. Esther, a 47 year old American Indian lesbian who was a member of MCC/NY and had been attending the church for over ten years, explained:

> *I think I'm blessed. Sometimes I can't express how blessed I feel. That God lives [at MCC/NY], God is at home here. I have a chance every Sunday to come here and be with God. Here I've a sense of connectedness and of belonging, and of being one with my community and with God at the same time. And what a double blessing that is for me.*

As can be seen from the above quotes, MCC/NY played an important role in enabling these gay and lesbian participants to integrate their sexual orientation and their religious beliefs. MCC/NY had marked itself in the minds of these participants as a Christian church that ministers to the gay and lesbian community of New York City by specifically addressing the spiritual needs of its gay and lesbian members and participants.

DISCUSSION

This study is an important step towards a more in-depth understanding of the psychology of gay and lesbian Christians. Not only does our work provide a more detailed examination of identity integration, but it is also one of the few in the homosexual and Christian identity literature that looks at both gay men and lesbians at the same time. The results presented here indicate that involvement at MCC/NY, for both gay men and lesbians, is an important means of identity integration for many of the participants in this study. Becoming involved in this particular gay-positive church was an invaluable tool for enabling many of the gay and lesbian members and participants at MCC/NY to achieve integration between their sexual orientation and their religious beliefs. For the gay and lesbian members and participants of MCC/NY, integration was found not only to be positively related to role involvement, worship service and activity/ministry attendance, membership, and length of time at MCC/NY, but to be positively related to being open about one's sexual orientation as well. This finding supports our definition of identity integration where individuals hold a positive religious identity, a positive gay identity, and do not currently feel conflict between the two. Yet this study also illustrated that for some of the research participants, integration could, and did, take place independent of their participation at MCC/NY. Reading relevant literature, developing self-acceptance, simply talking to other individuals, and even direct intervention by God were all mentioned as possible catalysts for the integration of homosexual and religious identities.

Although we did not predict that there would be gender differences, we found that not only did more lesbians than gay men report never experiencing conflict between their homosexual and Christian identities, but that lesbians also reported being fully integrated more often than their gay male counterparts. One possible explanation for these results is that while there was no difference in general level of involvement at MCC/NY, the lesbians in our sample attended more worship services and church ministries and activities than the gay men. They attended more, thus they reported more integration. It could also be the case that the lesbians at MCC/NY are more empowered than gay men by their experiences in the church. In our previous participant-observation study we documented how MCC/NY utilized gender-neutral language and was lead by a strong, popular lesbian pastor. While having such leadership and language usage is a boon to the entire congregation, such issues resonated strongly with the lesbian members and participants of the church (Rodriguez and Ouellette, 1999).

Also relevant is when Christian doctrine refers to the sin of homosexuality, it is primarily referring to gay men as opposed to lesbians (Carmody and Carmody, 1993; Nugent and Gramick, 1989). Of the six Biblical passages that make reference to homosexual behaviors (Genesis, 19:1-28; Leviticus, 18:22, 20:13; Romans, 1:26,27; I Corinthians, 6:9; I Timothy, 1:10), all refer in some way to "the abomination" of man-to-man sex, while only one (the passage from Romans) prohibits woman-to-woman sexual activity. Indeed, in our previous work we documented that the primary struggle some gay men at MCC/NY had with their Christian beliefs revolved around their becoming sexually active as gay men (Rodriguez and Ouellette, 2000).

It could be, however, that the lesbians at MCC/NY experienced their conflict with religion in a different way than their gay male counterparts. The lesbians at MCC/NY may struggle with religion not simply because they engage in sexual activity with other women, but because they are women. It is possible that for many of these women conflict between their homosexual and Christian identities is not as salient for them as it was for the gay men because of their previously experienced conflict with being a woman in a male-dominated Christian church. As they were growing up, many of the woman of MCC/NY had to battle the rigid, socially conservative gender roles promoted by fundamentalists Christians, as well as the patriarchal nature of many Christian churches where women were forbidden from becoming members of the clergy. This is an especially pertinent point for the lesbians presented in this study as a majority of them (56%) grew up in the Roman Catholic church. Of course, the gender differences uncovered here could simply be due to women having higher levels of religiosity than men. Social science research has shown that when compared to men, women tend to have a stronger personal faith, a stronger commitment to orthodox religious beliefs, and a higher level of involvement in worship services and other church activities (Thompson, 1991).

Future research on gender, sexual orientation, and religious beliefs is in order as existing literature also raises questions about this issue. Shokeid (1995) reported, as we do, that few women in his study of the gay and lesbian participants of a gay-positive Synagogue experienced conflict between their homosexual and Jewish identities. Mahaffy (1996), however, in her study of both church involved and non-involved lesbian Christians reported that 75% of her sample had in fact experienced some sort of conflict between their sexual orientation and their Christian religious beliefs. Only additional research using larger sample sizes and more multivariate types of analyses than we could conduct here will be able to shed additional light on the gender differences we uncovered.

Due to the richness and diversity of the integration stories told by the research participants (and due to the fact that each interview was one on one, with few of the interviews actually taking place at the church), we believe that the participants were not simply relying upon scripted integration stories operant in MCC/NY culture, but were instead reporting how they perceived the combination of their homosexual and religious identities at that particular moment in time - colored of course by all of their life experiences, including (but not limited to) those at MCC/NY. The variance found in membership status, church involvement, and length of time in the church also contributed to our perception that what was going on at this particular church could not be explained as simply a desire to conform to MCC/NY group norms.

While we measured the concept of integration using a cross-sectional design, many of the members and participants at MCC/NY talked about identity integration as a process with which they were still involved, or that they had just recently completed. The quantitative results showing that integrated individuals tended to be older and to have attended MCC/NY for a longer period of time supports this notion. Therefore, it might not be the construct of integration that should be of theoretical interest in studying gay and lesbian Christians, but the process of integration instead. Unfortunately, in hindsight we determined that the integration questions were not properly worded to address this process issue. Future research using a longitudinal design could address such questions as: What are the catalysts that initiate the integration process? What are the stages (if any) of the integration process? Is the integration process different for lesbians as opposed to gay men - if so how and why? Does church involvement play a different role in the integration process at different points in time?

An important point needs to be made here: Despite the traditional animosity that many Christian denominations show towards homosexual men and women, and despite the conflict that many gay and lesbian Christians have experienced between these two conflicting identities, many gay and lesbian Christians feel very strongly about their religious beliefs and about their homosexual identity. The feelings for each of these identities are so strong that they refuse to give up (or reject) either one. It may be that this process of refusing to sacrifice either identity, and a desire to be "out" as both gay or lesbian and Christian (rather than just simply to alleviate conflict between the two), could be the trigger that allows an individual to begin the integration process. A second observation is that the desire to merge together one's homosexual and religious identities does not necessarily follow a period of identity conflict between the two. Of the forty members and participants of MCC/NY presented here, twelve reported never having experienced conflict between their sexual orientation and their religious beliefs, and yet nine of these same individuals reported having fully integrated these two identities anyway. Future research is needed to better understand the phenomenon of achieving identity integration from both a history of conflict, and from a history lacking any conflict between one's Christian religion and one's sexual orientation.

Conclusion

Using both qualitative and quantitative analyses, the present study demonstrated the complexity of the identity integration process between homosexual and religious identities as pertaining to a select group of gay and lesbian Christians attending a specific gay-positive church. The issues of identity integration and involvement at MCC/NY presented here illustrated that not only is the Christian religion an important part of these individual's lives, but that many of these gay men and lesbians go through great lengths to live openly as both a homosexual and a Christian. Because we focused specifically on the members and participants of MCC/NY in the present study, it was not possible for us to address all of the potential issues that arise as gay and lesbian Christians attempt to achieve a state of integration between their homosexual and Christian identities, nor did we attempt to do so. Instead we focused on obtaining a more thorough understanding of the experiences of gay and lesbian Christians at MCC/NY, and exploring more deeply the concept of identity integration between these individual's sexual orientation and Christian religious beliefs.

Acknowledgements

This research was funded in part by a Grant-in-Aid from the Center for Lesbian and Gay Studies (CLAGS), and was based on a larger study "Gay and Lesbian Christians: Conflict and Integration Between Religious and Homosexual Identities" used to complete Mr. Rodriguez's second year project requirement in the Social/Personality Psychology Subprogram at The CUNY Graduate Center. The authors wish to thank the following individuals for their invaluable help with previous drafts of this paper: Michelle Fine, James Pratt, Tracey Revenson, Amy Schmidt, Dorinda Welle, and three anonymous reviewers. The authors also wish to thank their colleagues at The CUNY Graduate Center and, of course, the staff, leadership, members, and participants of the Metropolitan Community Church of New York.

References

Baumeister, Roy F., Jeremy P. Shapiro and Dianne M. Tice. 1985. Two kinds of identity crisis. *Journal of Personality* 53(3): 407-424.
Carmody, Denise and John Carmody. 1993. Homosexuality and Roman Catholicism. In *Homosexuality and world religions*, edited by Arlene Swindler, 135-148. Valley Forge, PA: Trinity.
Clark, J. Michael, Joanne C. Brown and Lorna M. Hochstein. 1990. Institutional religion and gay/lesbian oppression. *Marriage and Family Review* 14(3-4): 265-284.
Deaux, Kay. 1991. Social identities: Thoughts on structure and change. In *The relational self: Theoretical convergences in psychoanalysis and social psychology*, edited by R. C. Curtis, 77-93. New York, NY: Guilford.
Ellison, Marvin M. 1993. Homosexuality and Protestantism. In *Homosexuality and world religions*, edited by Arlene Swindler. 149-180. Valley Forge, PA: Trinity.
Englund, Michael E. 1991. *The Bible and homosexuality*. Gaithersburg, MD: Chi Rho Press.
Gonsiorek, John C. 1991. The empirical basis for the demise of the illness model of homosexuality. In *Homosexuality: Research implications for public policy*, edited by John C. Gonsiorek and James D. Weinrich, 115-136. Newbury Park, CA: Sage.
Greenberg, David F. and Marcia H. Bystryn. 1982. Christian intolerance of homosexuality. *American Journal of Sociology* 88(3): 515-548.
Jelen, Ted G. and Clyde Wilcox. 1991. Religious dogmatism among White Christians: Causes and effects. *Review of Religious Research* 33(1): 32-46.

Jorgensen, Danny L. 1989. *Participant observation: A methodology for human studies. Applied social research methods series, volume 15.* Newbury Park, CA: Sage.

Keysor, Charles W. 1979. *What you should know about homosexuality.* Grand Rapids, MI: Zondervan.

Lukenbill, Bernard W. 1998. Observations on the corporate culture of a gay and lesbian congregation. *Journal for the Scientific Study of Religion* 37(3): 440-452.

Mahaffy, Kimberly A. 1996. Cognitive dissonance and its resolution: A study of lesbian Christians. *Journal for the Scientific Study of Religion* 35(4): 392-402.

Marshall, Catherine and Gretchen B. Rossman. 1989. *Designing qualitative research.* Thousand Oaks, CA: Sage Publications.

Metropolitan Community Church of New York (MCC/NY). 1994. Membership handouts.

Melton, J. Gordon. 1991. *The Church speaks on: Homosexuality.* Detroit, MI: Gale Research.

Mishler, Elliot G. 1990. *Validation in inquiry-guided research: The role of exemplars in narrative studies.* Harvard Educational Review 60: 415-442.

Nugent, Robert and Jeannine Gramick. 1989. Homosexuality: Protestant, Catholic, and Jewish issues; A fishbone tale. In *Homosexuality and religion,* edited by Richard Hasbany, 7-46. New York, NY: Harrington Park.

Perry, Troy D. 1990. *Don't be afraid anymore: The story of Reverend Troy Perry and the Metropolitan Community Churches.* New York, NY: St. Martin's Press.

Piazza, Michael S. 1994. *Holy homosexuals: The truth about being gay or lesbian and Christian.* Dallas, TX: Sources of Hope.

Rodriguez, Eric M. 1997. Gay and lesbian Christians: Conflict and integration between religious and homosexual identities. Unpublished Master's Thesis. CUNY Graduate Center.

Rodriguez, Eric M., and Suzanne C. Ouellette. 1999. The Metropolitan Community Church of New York: A gay and lesbian community. *The Community Psychologist* 32(3): 24-29.

Rodriguez, Eric M., and Suzanne C. Ouellette. 2000. Religion and masculinity in Latino gay lives. In *Gay Masculinities,* edited by Peter Nardi, 101-129. Thousand Oaks, CA: Sage.

Rosenwald, George C. and Richard L. Ochberg. 1992. *Storied lives: The cultural politics of self-understanding.* New Haven, CT: Yale University.

Scanzoni, Letha and Virginia Ramey Mollenkott. 1978. *Is the homosexual my neighbor? Another Christian view.* San Francisco, CA: Harper Collins.

Shokeid, Moshe. 1995. *A gay synagogue in New York.* New York, NY: Columbia University.

Singer, Bennett L. and David Deschamps. 1994. *Gay and lesbian stats: A pocket guide of facts and figures.* New York, NY: Harper Collins.

Smith, John K. 1993. *Hermeneutics and qualitative inquiry. Concepts in qualitative research,* edited by David J. Flinders and Geoffrey E. Mills, 183-201. New York, NY: Teachers College.

Throckmorton, Warren. 1998. Efforts to modify sexual orientation: A review of the outcome literature and ethical issues. *Journal of Mental Health Counseling* 20(4): 283-304.

Thompson, Edward H. Jr. 1991. Beneath the status characteristic: Gender variations in religiousness. *Journal for the Scientific Study of Religion* 30(4): 333-347.

Thumma, Scott. 1991. Negotiating a religious identity: The case of the gay evangelical. *Sociological Analysis* 52(4): 333-347.

Wagner, Glenn, James Serafini, Judith Rabkin, Robert Remien and Janet Williams. 1994. Integration of one's religion and homosexuality: A weapon against internalized homophobia. *Journal of Homosexuality* 26(4): 91-110.

White, Mel. 1994. *Stranger at the gate: To be gay and Christian in America.* New York, NY: Simon and Shuster.

Identity experience among progressive gay Muslims in North America: A qualitative study within Al-Fatiha

OMAR MINWALLA[1], B. R. SIMON ROSSER[1], JAMIE FELDMAN[1], & CHRISTINE VARGA[2]

[1]Department of Family Practice and Community Health, University of Minnesota Medical School, Minneapolis, MN, USA, and [2]Catherine T. MacArthur Foundation, Chicago, IL, USA

Abstract

This qualitative study aims to document the identity experience of progressive gay Muslim men in a North American context. Six in-depth interviews, supplemented with participant observation, were conducted of gay Muslim men who attended an international conference for lesbian, gay, bisexual, transgendered, and questioning (LGBTQ) Muslims. For progressive gay Muslims such as these, a Muslim identity appears three-dimensional (religious, ethno-cultural, and color) when integrated with a gay identity. As a religious identity, gay Muslim's relationship to *Allah* (God) and a reinterpretation of the Qur'an and traditional condemnation of homosexuality appears necessary. As a cultural identity, East–West ethno-cultural differences that impact on homo-sociality and gay identity construction, marriage and the impact of coming out on the Eastern family and siblings emerged as critical issues. As a color identity, internalized racism, dating relationships and social dynamics within gay subculture as Muslims of color in a white dominant context appear key challenges.

Keywords: *Gay, Muslim, religious identity, homosexuality, Al-Fatiha*

Introduction

Homosexuality is a controversial issue in many religious traditions. Religious arguments are frequently used to justify punitive and rejecting policies regarding homosexuality and, in reaction, many men turn away from traditional religion on their journey toward homosexual identity formation (Rosser 1992, Rey 1997). For some gay persons, however, the experience of being exiled from family, church and mainstream society can give rise to a deeper sense of spirituality (Fortunato 1982, Rey 1997). Rosser (1992), for example, noted that in the early stages of coming out (i.e., acceptance of one's sexual orientation), rejection of religious tradition is common; but in later stages, when identity acceptance and pride are reached, the need to integrate one's homosexual and religious identities can become paramount. Fortunato (1982) asserts that sexuality and spirituality are intricately linked and that a positive appreciation of one's sexual orientation is essential to spiritual and psychosocial development.

Correspondence: B. R. Simon Rosser, Program in Human Sexuality, 1300 S. 2nd Street, Suite 180, Minneapolis, MN 55454, USA. E-mail: rosse001@umn.edu

ISSN 1369-1058 print/ISSN 1464-5351 online © 2005 Taylor & Francis Group Ltd
DOI: 10.1080/13691050412331321294

The intersection of gay and religious identities is an emerging area of social science research. Most prior work conducted in a North American context has focused on Jewish and Christian traditions (Nugent 1986, Cooper 1989, Kahn 1989, Dynes and Donaldson 1992, Rosser 1992) and a few studies of Native American Two-Spirit traditions (Williams 1986). Khan's (1997b) personal narrative, which describes the experience of being gay and Muslim in the context of Pakistan and Toronto, as well as selections in Ratti's (1993) collection of writings by queer South Asians, describe elements of the gay Muslim experience. However, aside from such poems, essays, and personal narratives (Dossani 1997, Lake 1999), we could find no academic studies addressing the experience of integrating a gay identity with a Muslim identity.

This does not mean that homosexuality is unknown to Islam. Despite contemporary perceptions that homosexuality is a Western phenomenon (AbuKhalil 1997), same-sex dynamics of many varieties are an integral part of Islamic history and culture (Murray and Roscoe 1997). In their seminal analysis of historical, anthropological, and literary studies and texts, Murray and Roscoe (1997) identified consistent patterns of Islamic homosexualities that can be traced over centuries. Furthermore, Schmitt and Sofer (1992) have documented sexuality and eroticism between males in contemporary Muslim societies. These accounts, however, are mainly travel writings by anglo/white sex tourists to Arab lands, and reveal the problematic scarcity of scholarly documentation of Muslim men describing their own experiences (AbuKhalil 1997). Although scholarship has also been devoted to examining homosexuality in Eastern and Western contexts (Harda 2001, Keogh et al. 2004), and to understanding the phenomenon of being men of color who have sex with men in white America (Gonzalez and Espin 1996, Jones and Hill 1996), we could find no studies specifically of homosexuality among Muslim men in the West, nor of gay Muslim identity integration.

In this ethnographic study, we explored the dual identity experience of a highly specific and relatively rare sample: gay Muslim men who are part of Al-Fatiha, a movement within Islam that validates the experience of and advocates for the voices of lesbian, gay, bisexual, transgendered and questioning Muslims. Al-Fatiha is an Internet-based organization that does not contain regional offices but consists of local chapters in various cities in the US and abroad that hold local, national and international gatherings. As background, we first summarize interpretations of homosexuality in Islam and the progressive movement of Al-Fatiha. This is followed by a description of our qualitative methods, results, and discussion of our findings.

Interpretations of homosexuality in Islam

Contemporary Muslim scholars contend that all humans are 'naturally' heterosexual; accordingly, homosexuality is considered a sinful and a perverse deviation from a person's true nature (Abu-Saud 1990). All Islamic schools of thought and jurisprudence consider homosexual acts to be unlawful, but each differ in terms of penalty—from severe punishment, including death (Hanabalites), to no punishment warranted (Hanafite).

The theological justification for Islam's rejection of homosexuality appears in the story of Lut ('Lot' in Hebrew scriptures), which is referenced seven times in the Qur'an (Wafer 1997), the sacred text of Islam. God sends two angels to Lut, who offers them shelter. The neighboring people surround Lut's house and demand that he release the visitors so that they 'might know them', suggesting gang rape of the men (de la Huerta 1999). The subsequent destruction of the people of Lut has traditionally been interpreted as a condemnation of their sexual practices, i.e., homosexuality (Jamal 2001). However, many contemporary Muslim, Jewish and Christian scholars and theologians question this traditional interpretation. Jamal (2001) asserts that in the Lut narrative, the people's sins are numerous and that the Qur'an does not state that destruction fell upon the people for a specific sin. Scholars suggest that inhospitality to strangers, for example—not homosexuality as we understand it today—was among the intended transgressions (Kahn 1989, Jung and Smith 1993, de la Huerta 1999). Furthermore, Jamal (2001) points out that the *hadith*, which are collections of sayings separate from the Qur'an but attributed to the prophet Muhammad, have connected the story of Lut and same-sex sexuality exclusively, thus influencing interpretations of the Qur'an. While the *hadith* are more explicit in condemning homosexual acts, the *hadith* themselves are controversial. Many liberal Muslims challenge their authenticity, whereas traditionalist orthodox Muslims contend that the *hadith* contain the authentic sayings of Muhammad.

As in other major religions, divergent understandings and interpretations of Islam's position toward homosexuality exist. Conservatives view the Qur'an as being 'very explicit in its condemnation of homosexuality, leaving scarcely any loophole for a theological accommodation of homosexuals in Islam' (Duran 1993: 181). Others who have analysed the issue via semantic analysis of Qur'anic passages, contend that same-sex indiscretions are not one 'of the most dangerous crimes' as is believed by certain traditional Muslims, and that the Qur'an's objections towards same-sex actions are on par with objections toward opposite-sex and non-sexual indiscretions alike (Jamal 2001). Furthermore, such analysis asserts that the Qur'an is not clear about the position of same-sex sexuality, especially about exclusive homosexuality or the queer lifestyle as is understood in the West (Jamal 2001). Liberals contend that the Qur'an neither mentions the type of punishment for homosexual acts, nor portrays a strongly negative attitude against such acts. Dossani (1997: 236) asserts, 'The roots of gay intolerance seem to be more sociological and cultural than religious'. Certainly, contemporary mainstream Islam officially condemns homosexuality, but there is a growing movement of progressive-minded Muslims, especially in the Western world, who view Islam as an evolving religion that must adapt to modern-day society (Al-Fatiha Foundation 2002). This study focused on the experience of progressive gay Muslims who are part of this emerging subculture and movement called Al-Fatiha.

The Al-Fatiha context

The Al-Fatiha Foundation is an international grassroots organization for Muslims and their friends who are lesbian, gay, bisexual, transgendered, or questioning their sexual orientation and gender identity (LGBTQ). The organization's stated goals are 'to provide a safe space and a forum for LGBTQ Muslims to address issues of common concern, share individual experiences and institutional resources ... [and] to support LGBTQ Muslims in reconciling their sexual orientation or gender identity with Islam' (Al-Fatiha Foundation 2002). The name Al-Fatiha is taken from the Qur'an and means 'the opening'. The foundation, a USA-based non-profit, non-governmental organization, was founded by a queer Muslim activist in 1997 via an internet list-serve and has since grown to include over 700 members, with chapters in the USA, Canada, the UK and South Africa. Its activities include establishing local support and discussion groups within the various chapters, holding meetings and providing outreach to other Muslims via the Internet.

Al-Fatiha's presence is controversial and even unimaginable to many Muslims. A British Muslim group, *Al-Muhajiroun*, issued a *fatwa* (an official legal opinion or decree issued by an Islamic religious leader) on July 16, 2001 stating:

> The very existence of Al-Fatiha is illegitimate and the members of this organization are apostates. Never will such an organization be tolerated in Islam, and never will the disease that it calls for be affiliated with a true Islamic society or individual. The Islamic ruling for such acts is death. It is a duty of all Muslims to prevent such evil conceptions from being voiced in the public or private arena. (Al-Muhajiroun 2002)

Notably, Al-Fatiha's goals extend beyond LBGTQ issues. Its members also seek to promote 'Islamic notions of social justice, peace, and tolerance, to bring all closer to a world that is free from prejudice, injustice, and discrimination' (Al-Fatiha Foundation 2002). Thus, Al-Fatiha can be considered part of a broader contemporary agenda and social movement referred to as Progressive Islam. Progressive Muslims define themselves in ways that are feminist, anti-racist, anti-violent and are committed to the greater movement for liberation and equality within the faith and within society.

Methodology

This study employed ethnographic methods to explore the dual identity experience of gay Muslim men who are part of Al-Fatiha. Data were collected primarily from in-depth interviews of six self-identified gay Muslim men who attended one of two annual Al-Fatiha conferences for LGBTQ Muslims, one held in the USA and one in Canada. This particular segment of gay Muslims can be defined as engaged in an active and conscious process of identity exploration within a Western-based, socio-political context. All participants had reached at least a sufficient level of identity development to self-identify as gay and as Muslim and were willing to talk about these issues with a researcher.

Setting

Data collection took place at the third and fourth annual international gatherings of Al-Fatiha, held respectively in May 2002 in Washington, DC and June 2003 in Toronto, Canada. Both conferences addressed issues of gender equality, reconciling identity and spirituality, the Qur'an and homosexuality, coming out, and social reform.

For the safety and security of conference attendees, all were required to register prior to the conference. The exact conference location was not publicly announced, but rather disclosed only to pre-registrants. Approximately 70–80 people attended each conference. Most lived in the USA or Canada, but others resided in the UK, Morocco, and Indonesia. The gender ratio was approximately 3:1 (M:F). Most of the 2002 attendees were between 18 and 30 years of age; the 2003 attendees were more heterogeneous in age.

Study participants

Following a brief announcement of the study, interested participants voluntarily contacted the senior author to be interviewed. Participants were limited to self-identified gay males (excluding females) to narrow the scope of analysis, given the small number of total participants. Of eight volunteers, six resulted in interviews due to time and scheduling constraints. Informed consent was obtained prior to the interview. Of the six gay Muslim men interviewed for this study, five were in attendance at the 2003 Al-Fatiha conference in Toronto; the sixth, the first pilot interviewee, attended the 2002 Washington meeting. All were residing in North America at the time of the interview. All spoke fluent English and were interviewed in English.

Study participants ranged in age from 18 to 48. Three were born in Pakistan and one in the Asian Peninsula; all four were raised from birth as Muslim. The fifth and sixth participants, African-American and Anglo-American, were both American-born and raised Christian before converting to Islam as adults.

Participant observation and creation of template interview

Participant observation was conducted at the Washington conference for the purpose of generating a template interview (subsequently utilized for the pilot interview). From both conferences, field notes were taken of the various lectures, workshops, small group discussions, panels and video presentations to validate the data from the interviews.

In-depth interviews

In-depth, semi-structured interviews were conducted to generate narratives that elucidated the internal psychological processes involved in identity experience. Each interview lasted approximately two hours. The constant comparative method of Glaser and Strauss (1967) was used, such that issues raised in earlier interviews were formulated as questions for subsequent interviewees.

The interviews were structured do as to explore five domains: (1) historical religious/spiritual identity formation, (2) current religious/spiritual identity, (3) historical sexual identity formation, (4) current sexual identity, and (5) the intersection of both religious/spiritual and sexual identities. These domains were examined using the following leading questions or statements: How did your Muslim identity develop over time? Describe your current Muslim identity. How did your sexual identity develop over time? Describe your current sexual identity. How do you experience both identities together now? Open-ended questioning permitted each participant to actively direct the interview. This maximized the ability of participants to describe his subjective identity experience.

Analysis

Interviews were transcribed and used as the primary data source. Emergent themes were generated for each transcript from a line-by-line microanalysis using a coding approach (open, axial, and selective) characteristic of grounded theory (Strauss and Corbin 1990). These themes were generated across domains and were not structured according to the interview domains. Thematic categories were compared across transcripts and refined. Relationships between categories were analysed. Field notes from the participant observation were used to further ground the results of the primary analysis, by confirming and giving validation to the emergent themes.

Results

Three prominent themes related to the intersection of gay and Muslim identities emerged from our dataset: (i) Religion, specifically one's relationship to Allah (e.g., as partner, betrayer) and different ways of coping with traditional interpretations of the Qur'an's condemnation of homosexuality; (ii) East-West ethno-cultural comparisons, such as homo-sociality and construction of a gay identity, marriage expectations and impact on sisters marriage potential; and (iii) color dynamics, such as internalized racism, validation theory and the impact on dating and sexuality. These themes are summarized below using illustrative examples from the interview transcripts.

Theme 1: Religion

Relationship to Allah. Among progressive Muslims of Al-Fatiha, relationship to Allah was a salient issue and frequent topic of discussion. Some men perceived Allah as a partner, turning to him for guidance and comfort as they sought to reconcile a perceived conflict between their gay and Muslim identities.

> Because I didn't know what was happening in the other [gay] world, I was struggling deep within myself ... I didn't talk about my struggle with my religious friends. Internally, I was sharing my struggles with God.

For some, turning to Allah and developing a stronger religious identity served to deflect heterosexist pressures and assumptions by family, including the pressure to marry. It was also used to distance oneself from and/or deny one's sexuality:

> When I was probably 17 or 18, I was struggling because I was attracted to men. There was a sense of guilt ... I found a mosque in the neighborhood that I would go to every Friday, and I found a group of people that were pretty religious, ... that was for me an escape from feeling guilty ... because I didn't know who to talk to about my feelings so I kind of concealed them, stored them away ... So, at 17, 18, and 19, that's when I was kind of very religious. Escape from socializing from my cousins, even with my parents, because each time we met as an extended family, there was talk about girls or marriage—it was just too much for me. That was a good escape from those pressures.

Others spoke of initially feeling betrayed or rejected by Allah as their understanding of their sexuality grew. In the earlier stages of sexual identity formation, some chose their sexual identity over their relationship with Allah or religious identity.

> As I started becoming more aware of Islam, learning about religion, I became more and more aware that it wasn't permissible to be queer and Muslim at the same time. There was a sense of conflict. I remember I would cry when I would read the Qur'an, because I felt that God is not going to permit me to be who I am, and I didn't feel that I'm doing anything wrong ... I feelthat this is instinctual to have desires towards men. I don't feel I have any control over it, but it's against Islam and I don't know what to do ... I felt betrayed, completely betrayed. It was like why is it that my brothers and sisters can look up to you, can pray to you, and you receive their prayers, but you won't receive mine? What is the difference between me and them?

Some comments reflected an intellectual challenging of Allah and a movement towards adopting more universal spiritual principles rather than religious doctrine.

I think that it's more important to be a good human being than a good Muslim, Christian or Jew. Because if being a good Muslim is not like being a good human being, than I'd rather be a good human being ... those things shouldn't conflict with each other.

The Qur'an and homosexuality. When discussing religion during the interviews, participants often made reference to the Qur'an, reflecting its prominence in Muslim discourse. Progressive gay Muslims have diverse ways of coping with the traditional view that the Qur'an explicitly prohibits homosexuality.

Some men questioned outright the idea that the Qur'an condemns homosexuality.

I can't see anything in the Qur'an that says homosexuality is wrong ... The Qur'an says many things are a sin ... All it needed to do was have one line that says it is a sin, and it doesn't do that.

Similarly, some participants disagreed with traditional interpretations of the story of Lut, or otherwise note the scholarly debate on this topic.

I came to the conclusion that the story was about straight men who were abusing or raping other men, like in war. It was shaming rituals. The intent behind it was rape, violence and oppression.

[In coming out to my family], I told them: I'm gay, and I'm Muslim, and I don't believe this is against Islam. I feel that there needs to be a reinterpretation of the story of Lut ... Some day I will prove to you that it's not the way everybody believes it is. The story has been misinterpreted.

Some men reinterpreted the authority of the Qur'an. They distinguished between the inspired word of Allah and the human aspects, such as Mohammed's transcription of it.

So much has happened through history, that I don't see how we can say that every single word in the Qur'an is true. And then there's the fact that he (Mohammed) is human ... The only thing that can't make mistakes is supposed to be God so how can it be possible that he didn't make a mistake?

Some men, early in the coming out process when they felt betrayed by Allah (see above), also reported distancing themselves from Islam and rejecting the Qur'an.

Realizing that there was a part of me that I cannot express to anybody and that it is forbidden in Islam and so it was a sense of not belonging to the religion because it does not allow for people like myself to exist or ... do what they find natural. My reaction to that initially was sorrow and despair, but after that I guess I made a choice that I'm just going to close the Qur'an and not ... read it anymore. There's no point. There's no room for me in it, in Islam. I cannot suppress my desires. If one of the things has to be jettison, then it's Islam. So, I just stopped participating.

Another way of coping expressed by participants was to reclaim the Qur'an, challenge misconceptions of it, and use it to help resolve the internal struggle between sexuality and religion.

You must read the Qur'an for yourself ... that's true surrender to Allah, and Allah will then respond to you. Until you do that, you really don't know ... We as homosexual men and women need to deal with the issue. If you're Muslim, then you need to read the Qur'an. By reading the Qur'an, you know it for yourself so that you will deal with your own demons.

Finally, the impact of cultural biases and cultural traditions on interpretations of the Qur'an were emphasized as distorting Islam the religion and the actual doctrine of the Qur'an.

> I think it's more of a social and traditional and cultural thing right now, rather than religion condemning it. Just the social values that people have.

Theme 2: East-West cultural comparisons

As an ethno-cultural identity, being Muslim and coming out as gay in the West are complicated by cultural differences. The key challenge is coming out and constructing a gay identity, and the impact of those decisions on intimacy, marriage, family and personal safety.

Homo-sociality and gay identity. Several men described East-West cultural differences in how homo-sociality is constructed, expressed, and understood.

> Here [in North America] I knew about gayness, but I think in Pakistan there's really no concept of gay. You can be really intimate with another guy. I saw a lot of men holding hands and doing all sorts of stuff. But it was never seen as being something wrong. Here, you can't kiss a man in public, can you? I mean, it's allowed, but who does it? In Pakistan, there's a lot of intimacy allowed between men. It's not labeled and has nothing to do with a gay identity. It's fine there.

> Because [in the East] you are allowed to say, 'You look beautiful'. Adoring another male's body is okay, not like here [in the West].

Coming out as gay. As migrants to the West (or North America), some men had some initial resistance to, and difficulty with, the Western process of constructing a gay identity. Particularly difficult was understanding and accepting the Western practice of putting labels on homo-social expression.

> In the West things are labeled. Here you have to pick a box that you're in. To get in that box, it took a long time to come to terms and say, 'Ok, I'm gay'. Eventually I came to terms, got in the box. They [in the East] don't really do this box. You're you.

Marriage expectations. In the East, marriage is seen as a cultural and traditional imperative, connected to reproductive obligation and the creation of family. Because of this, gay Muslim men can experience great pressure from their families to marry (heterosexually), regardless of their sexual orientation.

> They knew I think all along that I wasn't like the rest of the boys and that I was gay, but they had always assumed that even if that were the case I would eventually get married.

> There is pressure from the family to get married and all that. I keep pushing back saying, 'I'm busy or I'm at school or I just got this job and can't afford to get married'. It got to be so overwhelming … It came to a point where I broke down … I was in the psychiatric ward for almost 2 days.

Impact on sisters' marriage potential. Several participants described the impact of coming out on their sisters' potential for marriage.

[My parent's response when I came out was,] Fine, we believe you and that you can't change, but why do you have to advertise this in public? Why do you have to tell everybody about it? If you do this, who will marry your sisters? This is a common concern for any [Eastern] family. If our family is dishonored or socially stigmatized, we won't get any marriage for our daughters.

Physical danger of coming out. Finally, the danger of coming out as gay within a Muslim family should not be underestimated.

When I came out, I had a lot of threats of being killed. Right now [my family] don't think anyone else is going to find out. But if they thought I was going to come out and all, they probably would kick me out. There is so much I have to hide, but it's necessary for my safety. You have to make sure that being true to yourself doesn't mean getting killed.

Theme 3: Color dynamics

Five participants were not only gay and Muslim, but also men of color. For all six, color dynamics was a dominant topic. Their comments suggest that the color dimension of a gay Muslim identity plays an important role in social dynamics within the larger, predominantly white gay subculture of the West; internal dynamics (e.g., internalized racism, validation theory); dating preferences; sex behaviors.

Social dynamics within white gay culture. Participants clearly perceived that gay culture in the West is white dominated. Some expressed a sense that color can exclude them from gay social dynamics, or that their color identity is somehow separate from their gay identity.

White males were the majority, and so they were everywhere in high school. We lived in a small [Canadian] town. Everybody's white there. The attraction [to white males] was reinforced through media, through high school being surrounded by white guys, and also the gay model that most gay men are white ... I thought that if I'm going to be part of that, I'm obviously going to be with a white guy. I thought that there were no South Asian queers. I didn't see any images of South Asian queers. The few images that I did see were a few Black queers that were all drag queens, or they were negative images. I didn't feel attraction to them. Maybe my attraction [to white men] was what I was socialized with.

Several comments made reference to a developmental progression of identity formation. Simply identifying as gay, even if as part of white culture, may meet certain identity needs at an earlier stage of gay development. However, as one integrates a gay identity, color dynamics and the whiteness of gay culture can become an issue of concern.

Most of my friends at that point were white queer males. It didn't feel unnatural. I didn't feel like I was missing anything at that point, because I was happy to have finally met gay people. That was more than I could have asked for. But I think that slowly over time I became conscious of the fact that this is not enough. I can't go to gay bars and listen to Desai music. I can't go to Church Street [a street in the gay ghetto] and eat Indian food or anything like that. There is this separation between my South Asian identity and my queer identity.

I want to have my own community that recognizes my sexuality too, and I don't know if that's possible. I don't have a problem telling my gay friends of my Muslim background, but it somehow doesn't really feel like you connect on that dimension with them.

Some participants reported that within their circle of other gay men of color, there is a stigma associated with dating white gay men or assimilating into gay white culture.

> All my friends [gay Muslim South Asians] share the same perspective in terms of being very much against assimilating into gay white culture and going out with gay white men as well. They're all opposed to it. The other assumption would be from my South Asian friends that I'm uncomfortable being with other men of colour and therefore I've chosen to be with a white man because I'm internalizing my racism.

Internalized racism: The coconut syndrome. As people of color living in a white-dominant context, participants commented on the role of internalized racism—adopting negative and inferior thoughts and feelings around being a person of color. For example, one man described how he initially rejected his Islamic identity in attempts to assimilate into White culture. The following quote not only illustrates this process, but also demonstrates the evolution of accepting and integrating a color identity:

> I had this suspicion about Islam, and actually in my first years of living in England I stayed away from anything that had to do with Islam. I didn't even want Arabic friends. I was literally of the coconut syndrome, brown on the outside, white on the inside. I wanted to be white on the outside. I was in my late teens. That was almost 20 years ago. As you get older, you learn that these are your roots. Trying to pretend that you're not, it's so pathetic. It's about who you are and being honest and getting that peace within. (Arabic Muslim)

Validation theory. Some participants, mostly those of South Asian background, commented on the phenomenon whereby gay Muslim men of color are attracted exclusively to white men. This was framed in terms of validation theory, a term used to describe how being white—or to be in relation to that which is white—involves a gain in currency, power, or validation. Whereas some participants saw this phenomenon as highly prevalent, others questioned its legitimacy.

> One thing that I do know about gay Muslims is that they almost always have a white boyfriend. It's very funny at times. They live in that world of validation to have a white boyfriend. It's that thing where you don't find people who look like you attractive and ... What is that? ... I was at a banquet the other day, and I was with two other Muslims ... One of the guys here was the most WASP-ish, most average looking white guy. You have all these cute Arab boys and [they were] obsessed with this white guy ... I was at the dinner table with him, and he was one of the dullest people I have ever met. But he's white, and I guess you have more currency.

> If you and I had talked 5 years ago, I probably would have agreed with that [validation theory]. I don't know. Is there always a validation thing when you see a gay Muslim with a white guy? I don't have an answer to this. But now that I'm older, I just think let them be. Maybe they are genuinely in love with each other. That I don't know. But I know from my two friends here. I took them to a party, and they were gravitating towards every white guy, etc. That's their thing. I think in the gay culture it's pretty old, actually.

Color and dating. Many participants were conscious of color and its impact on the men they chose to date or with whom they form intimate relationships.

> I don't know what happened, but for some reason I completely lost my attraction to white men, completely ... it came to the point where I was exclusionary of white men. I thought that I

was expanding my horizons, but I was focused entirely on men of color in terms of physical attraction and in terms of my desire to be in a relationship. I think that part of it has to do with my level of analysis and also because I've become more critical ... more aware of society structures ... and so that internal awareness became internalized. It just sort of changed my orientation in terms of sexual attractions to men. I had to intellectualize it before feeling it as just a raw desire. It had to be something that I consciously chose.

Some discussed the comfort and affirmation they experience in dating or being in a relationship with another man of color of similar ethnic background.

We are both gay and Muslim, a Muslim Arabic couple. Dating someone from the Middle East was so comforting. It's kind of fun to just be who you are.

Being with him only affirmed all those feelings that this is what I feel comfortable with, being with a brown Muslim.

Boy-toy. Participants frequently mentioned the boy-toy concept—a derogatory term used to describe a (younger) man of color in a sexual or romantic partnership with a (older) white man.

A former roommate had told me that he had a crush on me ... I was 22 and he was 40, and I was really offended by it. It made me feel like I don't really want to be your South Asian boy-toy. I really don't want to be in a position where you have complete power over me. You're White, you're relatively well off economically, and you're older than me. You command a lot more authority in society in general and in the queer community. I don't want to be with somebody who has that much power, where there's that much of a power differential betweenmyself and my partner. When I told him, he was totally shocked. He was offended that I could think of our friendship in terms of power structures instead of being in a friend ship based on love that could blossom into a relationship. I was like, I have to think that way, I can't think otherwise. (South Asian Muslim).

Color and sexuality. A conscious awareness of power dynamics and color also impacted some men's sexuality.

Before this change happened [becoming more aware of color power differentials], I wasn't as conscious of the power differential within the sex that I would have. I would be a top and a bottom, I was totally versatile. After I became conscious of it, if I was at a club and I was horny that night and I met a guy and he happened to be white, I would not let him fuck me. I would only be the top. If it was person of color, I would allow myself to be the bottom or I would be versatile. With men of color, I was being more true to myself. I was allowing myself to be either/or, but with white men, I wouldn't allow them to fuck me. I was so conscious of that power differential that I didn't want to be placed in that position where somebody is dominating me within a sexual act position.

Discussion

This study is among the first to document gay Muslim identity experience of progressive gay Muslim men in a North American context. For this particular group, the gay-Muslim dual-identity appears to be at least three-dimensional, consisting of a religious identity, an ethno-cultural identity, and a color identity. On a more fundamental level, our research provides evidence that gay Muslims do indeed exist. When working with Muslim men struggling with sexual identity issues, health professionals should not assume that Islam and homosexuality are irreconcilable; our work demonstrates that at least some men experience their identities otherwise and are discovering ways to integrate them.

In this study, several men appeared to resolve their Muslim-gay identity conflicts by reinterpreting doctrine and emphasizing aspects of Islam that promote inclusion, peace, tolerance, and justice. Similar to research on gay religious identity development in Jews and Christians (Coyle and Rafalin 2000, Rosser 1992) a rejection of or distancing from religious identity often occurred in the earlier stages of sexual identity development. Others, however, strengthened their religious identity and turned to the Qur'an or Allah as they negotiated the initial stages of sexual identity development. Eventually, all reached a point of reclaiming and redefining their conceptualizations of what it means to be Muslim in a religious context.

For Muslim men, coming out as gay in North America can be complicated by cultural differences. For example, given the collective identity of family in many Eastern cultures (Abu-Saud 1990, Khan 1997a, b, Daneshpour 1998) it is important to consider the impact of coming out on the larger family unit, particularly female siblings and their potential for marriage. Additionally, respondents indicated that in Muslim society, same-sex attraction does not by definition nullify the cultural imperative for heterosexual marriage. This has been documented previously in South Asian and Eastern cultures, where the pattern of being heterosexually married and having sex with men is not necessarily contradictory (Khan 1997a, Seabrook, 1999).

Participants described their Eastern cultures as more permissive of homo-sociality, particularly expressions of physical intimacy and emotional closeness. In many Eastern cultures, intimate expressions between men are not constructed as sexual or feminine. Therefore, these behaviors are not internalized or perceived in a context of homosexuality, shame or stigma as they are in the West. Instead they are often internalized as part of cultural brotherhood, deep friendship and masculinity, similar to the Mediterranean pedagogical-Socratic love ideal (Khan 1997a). Notably, in our study, Muslim men from Eastern cultures appeared to have a heightened awareness that the process of constructing of a gay identity—that is, constructing homo-social expression into an internal and social identity—is more of a Western process. This may be why some respondents seemed to struggle with and resist the construct of gay, perceiving it as a confining label or a box—a resistance to cultural assimilation.

It is an interesting paradox that Eastern cultures, while more permissive of homo-social expression than Western cultures, are more repressive and hostile to a gay identity. Eastern culture reacts to and restricts one's freedom to assert a sexual and social identity based on same-sex attraction, yet it allows men more freedom to express intimacy and love, particularly when it comes to such expression in public and without shame. This challenges the Eurocentric assumption that Eastern culture is more repressive, restrictive and hostile towards intimacy between men. This also pushes the understanding of gay liberation in the West and how rights are constructed. For example, if the West were to emulate the permissiveness of Eastern culture, it is interesting to consider the possibility of gay rights including the right of men, both heterosexual and homosexual, to hold hands or put their arms around each other in public, without shame and without the threat of persecution or sexualization.

Being men of color in North America represents a third dimension of this multifaceted Muslim identity experience. As a color identity, being Muslim appeared most salient to the gay identity process in terms of gay dating relationships as well as social and political dynamics related to the gay community. These men were conscious of living in a white-dominant culture and the power dynamics inherent in this context. Thus, gay Muslims of color experience similar dynamics of exclusion, marginalization and oppression to that documented among other gay men of color (Gonzalez and Epsin 1996, Jones and Hill 1996). For example, terms such as coconut syndrome and boy-toy resonate with terms documented among African-American gay men such as snow queen or dinge queen (Hemphill and Beam 1991) and among East Asians such as rice queen or sticky rice (Cho 1998). Such terms illustrate the common challenges experienced around racism, dating and identity development by gay men of color in a North American context.

At the same time, cultural variables may influence gay Muslim men in the North American context in ways that are unique to them. For example, the unease reported among participants with being a boy-toy for an older Anglo-American partner may not simply be due to color power differentials that exist in North America, but also to the pervasiveness of pederastic relationships as the predominant idiom of male-male sexual relations in Muslim societies in which the older uses the younger, who then often becomes identified and stigmatized as fair game to be penetrated by others (Murray 1997). Furthermore, this study is among the first to note how color impacts sexual behaviors such as being the active or passive partner during anal intercourse for gay Muslims. This distinction also has prominent significance among Muslim same-sex dynamics, both historically and in Muslim countries today (Schmitt and Sofer 1992, Tapinc 1992, Murray and Roscoe 1997). Men who have sex with men in a Muslim context typically distinguish themselves as either active or passive, with the active partner experiencing little if any stigma, while the passive partner is considered effeminate and highly stigmatized (Schmitt and Sofer 1992, Tapinc 1992). In fact, based on this distinction of active and passive, both gender and sexual identity is formulated—the active understood as masculine and heterosexual, while the passive is understood as feminine and homosexual (Tapinc 1992). Thus, status differentiation based on age and sexual position have implications for gay Muslim men who are aware of power differentials and who contend with pressures to assimilate, internalized racism and exclusion within the gay community. Future research may further explore how color dynamics influence sexual behavior and psychology among gay Muslim men of color who live in a white-dominant context and within the predominantly white gay subculture.

This study is limited in that it only explores the experience of progressive gay Muslim men in the context of Al-Fatiha. Because Al-Fatiha is a new and politically liberal movement, there are most likely significant differences between these men and other Muslim men who either do not openly identify as gay or have rejected their Muslim identity. Even within the context of Al-Fatiha, there was an undercurrent of suspicion among some men when they learned of this research being conducted. This further limited our sample, in that it includes only those Al-Fatiha participants who were willing to be interviewed. A second limitation is the small sample size. All findings should be considered preliminary in understanding the experience of this group of men, and clearly more research is needed to confirm or refute our findings. Finally, this research is limited to gay men. Future research should consider the experience of female, lesbian, bisexual and transgender individuals within the context of progressive Islam in North America and beyond. Other promising areas of study include examining the impact of the September 11, 2001 terrorist attacks in the USA on gay Muslims in North America, which was a predominant topic of discussion at the conferences, and the perceptions of HIV/AIDS risk among gay Muslim men, HIV prevention strategies for this population, and indeed, the physical, mental and emotional needs of this population.

Acknowledgements

The authors would like to acknowledge and thank Al-Fatiha Foundation, and particularly its founder, for permission to conduct the research; the participants for having the willingness and courage to participate; Pricilla Palm for her transcription services; and Anne Marie Weber-Main for her critical review and editing of manuscript drafts.

References

Al-Fatiha Foundation, Inc. 2002. Accessed, available at: http://www.al-fatiha.net.
Al-Muhajiroun, Accessed 28 March 2002, available at: http://www.religioustolerance.org/hom_isla.htm
AbuKhalil, A. (1997) Gender boundaries and sexual categories in the Arab world. *Feminist Issues*, 15, 91–104.
Abu-Saud, M. (1990) *Concept of Islam* (Indianapolis, IN: American Trust Publications).
Cho, S. (1998) *Rice: Explorations into Gay Asian Culture and Politics* (Toronto: Queer Press).
Cooper, A. (1989) No longer invisible: Gay and lesbian Jews build a movement. *Journal of Homosexuality*, 18, 83–94.
Coyle, A. and Rafalin, D. (2000) Jewish gay men's accounts of negotiating cultural, religious, and sexual identity: A qualitative study. *Journal of Psychology and Human Sexuality*, 12(4), 21–48.
Daneshpour, M. (1998) Muslim families and family therapy. *Journal of Marital and Family Therapy*, 24, 355–368.
De la Huerta, C. (1999) *Coming Out Spiritually: The next step* (New York: Putnam).
Dossani, S. (1997) Being Gay and Muslim. In G. Comstock and S. Henking (eds.) *Que(e)rying Religion: A Critical Anthology* (New York: Continuum), pp. 236–237.
Duran, K. (1993) Homosexuality and Islam. In A. Swidler (ed.) *Homosexuality and World Religions* (Valley Forge, PA: Trinity Press International), pp. 181–197.
Dynes, W. and Donaldson, S. (1992) *Homosexuality and Religion and Philosophy* (New York: Garland Publishing).
Fortunato, J.E. (1982) *Embracing the Exile* (New York: The Seabury Press).
Glaser, B.G. and Strauss, A.L. (1967) *The Discovery of Grounded Theory: strategies for qualitative research* (London: Weidenfield & Nicolson).
Gonzalez, F.J. and Espin, O.M. (1996) Latino Men, Latina Women, and Homosexuality. In R.P. Cabaj and T.S. Stein (eds.) *Textbook of Homosexuality and Mental Health* (Arlington, VA: American Psychiatric Association), pp. 549–561.
Harda, M. (2001) Japanese male gay and bisexual identity. *Journal of Homosexuality*, 42, 77–100.
Hemphill, E. and Beam, J. (1991) *Brother to Brother* (Boston, MA: Alyson Publications, Inc).
Jamal, A. (2001) The story of Lot and the Qur'an's perceptions of the morality of same-sex sexuality. *Journal of Homosexuality*, 41, 1–88.
Jones, B.E. and Hill, J. (1996) African American Lesbians, Gay Men, and Bisexuals. In R.P. Cabaj and T.S. Stein (eds.) *Textbook of Homosexuality and Mental Health* (Arlington, VA: American Psychiatric Association), pp. 549–561.
Jung, P. and Smith, R. (1993) *Heterosexism, an Ethical Challenge* (New York: State University of New York).
Keogh, P., Dodds, C. and Henderson, L. (2004) *Migrant gay men: Redefining community, restoring identity* (London: Sigma Research).
Kahn, R.Y.H. (1989) Judaism and homosexuality: the traditionalist/progressive debate. *Journal of Homosexuality*, 18, 47–82.
Khan, B. (1997a) Not-so-Gay Life in Pakistan in the 1980s and 1990s. In S.O. Murray and W. Roscoe (eds.) *Islamic Homosexualites: Culture, History, and Literature* (New York: New York University), pp. 275–296.
Khan, B. (1997b) *Sex, Longing, and Belonging: A Gay Muslim's Quest for Love and Meaning* (Bangkok: Bua Luang Books).
Lake, C. (ed.) (1999) *Recreations: Religion and Spirituality in the Lives of Queer People* (Toronto: Queer Press).
Murray, S.O. (1997) The Will Not to Know: Islamic accommodations of male homosexuality. In S.O. Murray and W. Roscoe (eds.) *Islamic Homosexualites: Culture, History, and Literature* (New York: New York University), pp. 14–54.
Murray, S.O. and Roscoe, W. (1997) *Islamic Homosexualites: Culture, History, and Literature* (New York: New York University).

Nugent, R. (ed.) (1986) *A Challenge to Love: Gay and Lesbian Catholics in the Church* (New York: Crossroad).
Ratti, R. (ed.) (1993) *A Lotus of Another Color: An Unfolding of the South Asian Gay and Lesbian Experience* (Boston, MA: Alyson).
Rey, P.G. (1997) *Homosexuality, identity formation, gay spirituality and family life-cycle*, dissertation abstracts international: AAT 9835945.
Rosser, S.B.R. (1992) *Gay Catholics Down Under* (Westport, CN: Praeger).
Schmitt, A. and Sofer, J. (eds.) (1992) *Sexuality and Eroticism Among Males in Moslem Societies* (New York: The Haworth Press).
Seabrook, J. (1999) *Love in a Different Climate: Men who have sex with men in India* (London: Verso).
Strauss, A. and Corbin, J. (1990) *Basics of Qualitative Research: Grounded theory procedures and techniques* (Newbury Park, CA: Sage).
Tapinc, H. (1992) Masculinity, Femininity and Turkish Male Homosexuality. In K. Plummer (ed.) *Modern Homosexualities* (London: Routledge), pp. 39–49.
Wafer, J. (1997) Muhammad and Male Homosexuality. In S.O. Murray and W. Roscoe (eds.) *Islamic Homosexualites: Culture, History, and Literature* (New York: New York University), pp. 87–96.
Williams, W.L. (1986) *The Spirit and the Flesh: Sexual Diversity in American Indian Culture* (Boston, MA: Beacon Press), pp. 17–30.

Résumé

Cette étude qualitative vise à documenter l'expérience identitaire d'hommes musulmans, gays et progressistes, dans un contexte nord américain. Six entretiens en profondeurs, complétés par de l'observation participante, ont été réalisés avec des hommes musulmans et gays qui assistaient à une conférence internationale rassemblant des lesbiennes, des gays, des bisexuel(le)s, des transgenres, musulmans et en questionnement sur leur religion. Il semble que pour les musulmans gays et progressistes comme les participants à cette conférence, l'identité musulmane soit tridimensionnelle (religieuse, ethno-culturelle et raciale) lorsqu'elle est intégrée à une identité gay. Dans sa dimension religieuse, cette identité musulmane fait que la relation des musulmans gays à Allah (Dieu), ainsi qu'une ré-interprétation du Coran et de la condamnation traditionnelle de l'homosexualité semblent nécessaires. Dans sa dimension culturelle, elle fait émerger les différences ehtno-culturelles entre Est et Ouest qui ont un impact sur l'homo-socialité et sur la construction de l'identité gay, le mariage et l'impact du *coming out* sur la famille orientale et les frères et sœurs, en tant que problèmes critiques. Dans sa dimension raciale, le racisme intériorisé, les relations amoureuses et la dynamique sociale au sein de la sous-culture gay, semblent représenter des défis majeurs pour les musulmans de couleur dans un contexte d'homme blanc dominant.

Resumen

La finalidad de este estudio cualitativo es documentar las diferentes identidades experimentadas por homosexuales musulmanes progresistas en la sociedad norteamericana. Se llevaron a cabo seis entrevistas exhaustivas, complementadas con la observación de participantes, con musulmanes homosexuales que asistieron a una conferencia internacional para la comunidad de lesbianas, gays, bisexuales, transexuales y personas con dudas (LGBTQ) de religión musulmana. Para estos homosexuales musulmanes progresistas, aparece una identidad musulmana tridimensional (religiosa, etnocultural y de color) cuando se integra en una identidad homosexual. Como identidad religiosa, parecen necesarias las relaciones de los homosexuales musulmanes con Allah (Dios) y una reinterpretación de la Qur'an y la tradicional condena de la homosexualidad. Como identidad cultural, parece ser que los retos clave son los problemas críticos que surgen en las diferencias etnoculturales este-oeste que influyen en la homosocialidad y la construcción de la identidad *gay*, el matrimonio y el impacto de salir del armario en una familia de zonas del este, especialmente en el caso de los hermanos. Como identidad del color de la piel, los desafíos básicos son el racismo interiorizado, relaciones de parejas y las dinámicas sociales en la subcultura homosexual de musulmanes de color en una sociedad predominantemente blanca.

All material is printed with copyright permission.

COURSEWARE IS NON-RETURNABLE

Sociology 3U0
3703

Custom Publishing

Everything in one book.

  The Campus Store and Media Production Services are proud partners in the production of this custom publication. All material is printed with copyright permission.

This custom publication is non-returnable